ELECTROMOTIVE FORCE AND MEASUREMENT IN SEVERAL SYSTEMS

Edited by **Sadik Kara**

Electromotive Force and Measurement in Several Systems
Edited by Sadik Kara

Published by InTech
Janeza Trdine 9, 51000 Rijeka, Croatia

Publishing Process Manager Oliver Kurelic

Technical Editor Teodora Smiljanic

Cover Designer InTech Design Team

Additional hard copies can be obtained from orders@intechopen.com

Electromotive Force and Measurement in Several Systems, Edited by Sadik Kara
p. cm.
ISBN 978-953-307-728-4

For EU product safety concerns:
IN TECH d.o.o., Prolaz Marije Krucifikse Kozulić 3, 51000 Rijeka, Croatia,
info@intechopen.com or visit our website at intechopen.com.

Meet the editor

Professor Sadık Kara is a Professor of Electrical and Electronics Engineering Department and Director of the Institute of Biomedical Engineering at Fatih University, Istanbul, Turkey. He received his Diploma in Electrical Engineering and his Doctorate in Biomedical Engineering Science, in 1991 and 1995, respectively. He worked as an engineer at Turkish Electricity Association, Textile Factory, and Military Maintenance Centre. From 1995-2008 he was a researcher and lecturer at Department of Electrical and Electronics Engineering, Erciyes University. Since 2008 he has been at Fatih University. He was also the Editor in Chief for the Turkish Journal of Electrical Engineering & Computer Sciences. He is a recipient of the Outstanding Research Award of several foundations and an author of over 200 publications, including 60 research papers in international refereed journals, conferences, workshop proceedings and printed books. His research interests include Electrical Devices, Electronics, Medical Electronics, Biomedical Instrumentation and Measurements, Biomedical Signal and Image Processing, Neuropsychiatric and Musical Measurements.

Contents

Preface

These days, it is impossible to think of an area of knowledge that can keep on developing without a collaboration in interdisciplinary fields.

Electromotive force is a type of energy per unit of electric charge that is converted reversibly from chemical, mechanical, or other forms of energy into electrical energy by a dynamo, battery, electrochemical cells, thermoelectric devices, solar cells, transformers, and so on.

The theory of electromotive Force is currently one of the most important research topics within the electrical engineering domain. Recently, the development of high throughput techniques, such as in solar energy, photocatalysis, High-Temperature Systems and solid-state electrolyte have increased the need of Application of Electromotive Force and Measurement.

This book presents some theoretical issues and a variety of Electromotive force applications. For better understanding, the chapters were grouped into two sections (Theoretical issues of Electromotive Force and Application of Electromotive Force). In Theoretical issues of Electromotive Force, the chapters are more oriented towards theoretical issues. Application of Electromotive Force consists of application-oriented chapters that report Application of Electromotive Force Measurements in several fields such as Nuclear Systems Using Lead Alloys, High-Temperature Systems, Mixed Solvents and Coils. Chapters in both sections are stand-alone and readers can commence from any chapter of interest to them. It is anticipated that this book will raise awareness about the Electromotive Force application field and help in setting up new research areas in Electromotive Force.

In conclusion, the main objective of this book was to present a broad range of well worked out, recent application studies as well as theoretical contributions in the field of applications of Electromotive Force.

As the editor, I would like to thank all the authors of this book, reviewers and the editorial staff of InTech Open Access Publisher for the successful completion of this ambitious objective. Without their contribution, it would have been impossible to publish a book of this quality and help in the development of this issue.

Sadık Kara
Fatih University
Istanbul,
Turkey

Part 1

Theoretical Issues of Electromotive Force

1

Quantum Theory of Thermoelectric Power (Seebeck Coefficient)

Shigeji Fujita[1] and Akira Suzuki[2]
[1]*Department of Physics, University at Buffalo, SUNY, Buffalo, NY*
[2]*Department of Physics, Faculty of Science, Tokyo University of Science, Shinjyuku-ku, Tokyo*
[1]*USA*
[2]*Japan*

1. Introduction

When a metallic bar is subjected to a voltage (V) or a temperature (T) difference, an electric current is generated. For small voltage and temperature gradients we may assume a linear relation between the electric current density $\boldsymbol{j}$ and the gradients:

$$\boldsymbol{j} = \sigma(-\boldsymbol{\nabla} V) + A(-\boldsymbol{\nabla} T) = \sigma \boldsymbol{E} - A\boldsymbol{\nabla} T, \tag{1.1}$$

where $\boldsymbol{E} \equiv -\boldsymbol{\nabla} V$ is the electric field and σ the conductivity. If the ends of the conducting bar are maintained at different temperatures, no electric current flows. Thus from Eq. (1.1), we obtain

$$\sigma \boldsymbol{E}_{\rm S} - A\boldsymbol{\nabla} T = 0, \tag{1.2}$$

where $\boldsymbol{E}_{\rm S}$ is the field generated by the thermal electromotive force (emf). The *Seebeck coefficient (thermoelectric power)* S is defined through

$$\boldsymbol{E}_{\rm S} = S\boldsymbol{\nabla} T, \quad S \equiv A/\sigma. \tag{1.3}$$

The conductivity σ is positive, but the Seebeck coefficient S can be positive or negative. We see that in Fig. 1, the measured Seebeck coefficient S in Al at high temperatures (400 – 670 °C) is negative, while the S in noble metals (Cu, Ag, Au) are positive (Rossiter & Bass, 1994). Based on the classical statistical idea that different temperatures generate different electron drift velocities, we obtain

$$S = -\frac{c_{\rm V}}{3ne}, \tag{1.4}$$

where $c_{\rm V}$ is the heat capacity per unit volume and n the electron density. A brief derivation of Eq. (1.4) is given in Appendix. Setting $c_{\rm V}$ equal to $3nk_{\rm B}/2$, we obtain the *classical formula* for thermopower:

$$S_{\rm classical} = -\frac{k_{\rm B}}{2e} = -0.43 \times 10^{-4}\,{\rm VK}^{-1} = -43\,\mu{\rm VK}^{-1}. \tag{1.5}$$

Observed Seebeck coefficients in metals at room temperature are of the order of microvolts per degree (see Fig. 1), a factor of 10 smaller than $S_{\rm classical}$. If we introduce the Fermi-statistically

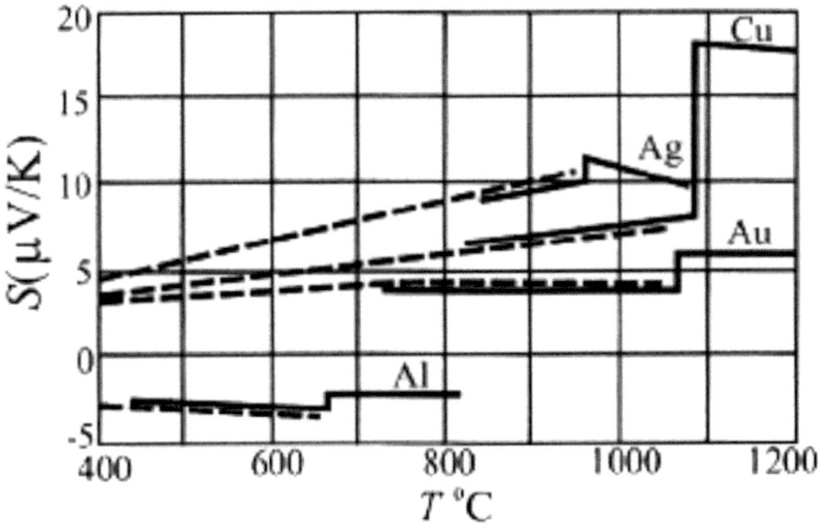

Fig. 1. High temperature Seebeck coefficients above 400 °C for Ag, Al, Au, and Cu. The solid and dashed lines represent two experimental data sets. Taken from Ref. (Rossiter & Bass, 1994).

computed specific heat

$$c_{\mathbb{V}} = \frac{1}{2}\pi^2 n k_{\mathrm{B}}(T/T_{\mathrm{F}}), \tag{1.6}$$

where T_{F} $(\equiv \varepsilon_{\mathrm{F}}/k_{\mathrm{B}})$ is the Fermi temperature in Eq. (1.4), we obtain

$$S_{\text{semi quantum}} = -\frac{\pi}{6}\frac{k_{\mathrm{B}}}{e}\left(\frac{k_{\mathrm{B}}T}{\varepsilon_{\mathrm{F}}}\right), \tag{1.7}$$

which is often quoted in materials handbook (Rossiter & Bass, 1994). Formula (1.7) remedies the difficulty with respect to magnitude. But the correct theory must explain the two possible signs of S besides the magnitude.

Fujita, Ho and Okamura (Fujita et al., 1989) developed a quantum theory of the Seebeck coefficient. We follow this theory and explain the sign and the T-dependence of the Seebeck coefficient. See Section 3.

2. Quantum theory

We assume that the carriers are conduction electrons ("electron", "hole") with charge q ($-e$ for "electron", $+e$ for "hole") and effective mass m^*. Assuming a one-component system, the Drude conductivity σ is given by

$$\sigma = \frac{nq^2\tau}{m^*}, \tag{2.1}$$

where n is the carrier density and τ the mean free time. Note that σ is always positive irrespective of whether $q = -e$ or $+e$. The Fermi distribution function f is

$$f(\varepsilon; T, \mu) = \frac{1}{e^{(\varepsilon-\mu)/k_{\mathrm{B}}T} + 1}, \tag{2.2}$$

where μ is the chemical potential whose value at 0 K equals the Fermi energy ε_F. The voltage difference $\Delta V = LE$, with L being the sample length, generates the chemical potential difference $\Delta\mu$, the change in f, and consequently, the electric current. Similarly, the temperature difference ΔT generates the change in f and the current.
At 0 K the Fermi surface is sharp and there are no conduction electrons. At a finite T, "electrons" ("holes") are thermally excited near the Fermi surface if the curvature of the surface is negative (positive) (see Figs. 2 and 3). We assume a high Fermi degeneracy:

$$T_F \gg T. \tag{2.3}$$

Consider first the case of "electrons". The number of thermally excited "electrons", N_x, having energies greater than the Fermi energy ε_F is defined and calculated as

$$N_x = \int_{\varepsilon_F}^{\infty} d\varepsilon\, \mathcal{N}(\varepsilon) \frac{1}{e^{(\varepsilon-\mu)/k_B T}+1} = \mathcal{N}_0 \int_{\varepsilon_F}^{\infty} d\varepsilon \frac{1}{e^{(\varepsilon-\mu)/k_B T}+1}$$

$$= -\mathcal{N}_0\,(k_B T)\left[\ln[1+e^{-(\varepsilon-\mu)/k_B T}]\right]_{\varepsilon_F}^{\infty} \cong \ln 2\; k_B T \mathcal{N}_0, \quad \mathcal{N}_0 = \mathcal{N}(\varepsilon_F), \tag{2.4}$$

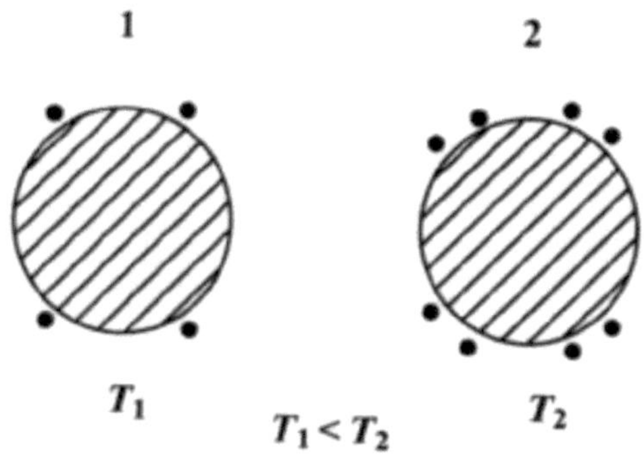

Fig. 2. More "electrons" (dots) are excited at the high temperature end: $T_2 > T_1$. "Electrons" diffuse from 2 to 1.

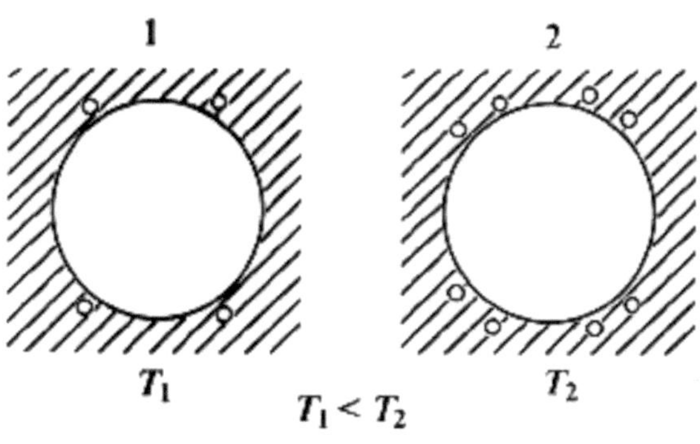

Fig. 3. More "holes" (open circles) are excited at the high temperature end: $T_2 > T_1$. "Holes" diffuse from 2 to 1.

where $\mathcal{N}(\varepsilon)$ is the density of states. The excited "electron" density $n \equiv N_x/\mathbb{V}$ is higher at the high-temperature end, and the particle current runs from the high- to the low-temperature end. This means that the electric current runs towards (away from) the high-temperature end in an "electron" ("hole")-rich material. After using Eqs. (1.3) and (2.4), we obtain

$$S < 0 \quad \text{for} \quad \text{"electrons"}, \qquad S > 0 \quad \text{for} \quad \text{"holes"}. \tag{2.5}$$

The Seebeck current arises from the thermal diffusion. We assume Fick's law:

$$\boldsymbol{j} = q\boldsymbol{j}_{\text{particle}} = -qD\boldsymbol{\nabla} n, \tag{2.6}$$

where D is the *diffusion constant*, which is computed from the standard formula:

$$D = \frac{1}{\mathrm{d}} vl = \frac{1}{\mathrm{d}} v_{\mathrm{F}}^2 \tau, \quad v = v_{\mathrm{F}}, \quad l = v\tau, \tag{2.7}$$

where d is the dimension. The density gradient $\boldsymbol{\nabla} n$ is generated by the temperature gradient $\boldsymbol{\nabla} T$ and is given by

$$\boldsymbol{\nabla} n = \frac{\ln 2}{\mathbb{V}\mathrm{d}} k_{\mathrm{B}} \mathcal{N}_0 \boldsymbol{\nabla} T, \tag{2.8}$$

where Eq. (2.4) is used. Using the last three equations and Eq. (1.1), we obtain

$$A = \frac{\ln 2}{\mathbb{V}} q v_{\mathrm{F}}^2 k_{\mathrm{B}} \mathcal{N}_0 \tau. \tag{2.9}$$

Using Eqs. (1.3), (2.1), and (2.9), we obtain

$$S = \frac{A}{\sigma} = \frac{2\ln 2}{\mathrm{d}} \left(\frac{1}{qn}\right) \varepsilon_{\mathrm{F}} k_{\mathrm{B}} \frac{\mathcal{N}_0}{\mathbb{V}}. \tag{2.10}$$

The relaxation time τ cancels out from the numerator and denominator.

The derivation of our formula [Eq. (2.10)] for the Seebeck coefficient S was based on the idea that the Seebeck emf arises from the thermal diffusion. We used the high Fermi degeneracy condition (2.3): $T_{\mathrm{F}} \gg T$. The relative errors due to this approximation *and* due to the neglect of the T-dependence of μ are both of the order $(k_{\mathrm{B}}T/\varepsilon_{\mathrm{F}})^2$. Formula (2.10) can be negative or positive, while the materials handbook formula (1.7) has the negative sign. The average speed v for highly degenerate electrons is equal to the Fermi velocity v_{F} (independent of T). Hence, semi-classical Equations (1.4) through (1.6) break down. In Ashcroft and Mermin's (AM) book (Ashcroft & Mermin, 1976), the origin of a positive S in terms of a mass tensor $M = \{m_{ij}\}$ is discussed. This tensor M is real and symmetric, and hence, it can be characterized by the principal masses $\{m_j\}$. Formula for S obtained by AM [Eq. (13.62) in Ref. (Ashcroft & Mermin, 1976)] can be positive or negative but is hard to apply in practice. In contrast our formula (2.10) can be applied straightforwardly. Besides our formula for a one-carrier system is T-independent, while the AM formula is linear in T.

Formula (2.10) is remarkably similar to the standard formula for the Hall coefficient:

$$R_{\mathrm{H}} = (qn)^{-1}. \tag{2.11}$$

Both Seebeck and Hall coefficients are inversely proportional to charge q, and hence, they give important information about the carrier charge sign. In fact the measurement of the

thermopower of a semiconductor can be used to see if the conductor is n-type or p-type (with no magnetic measurements). If only one kind of carrier exists in a conductor, then the Seebeck and Hall coefficients must have the same sign as observed in alkali metals.
Let us consider the electric current caused by a voltage difference. The current is generated by the electric force that acts on *all* electrons. The electron's response depends on its mass m^*. The density (n) dependence of σ can be understood by examining the current-carrying steady state in Fig. 4 (b). The electric field $\boldsymbol{E}$ displaces the electron distribution by a small amount $\hbar^{-1}qE\tau$ from the equilibrium distribution in Fig. 4(a). Since all the conduction electron are

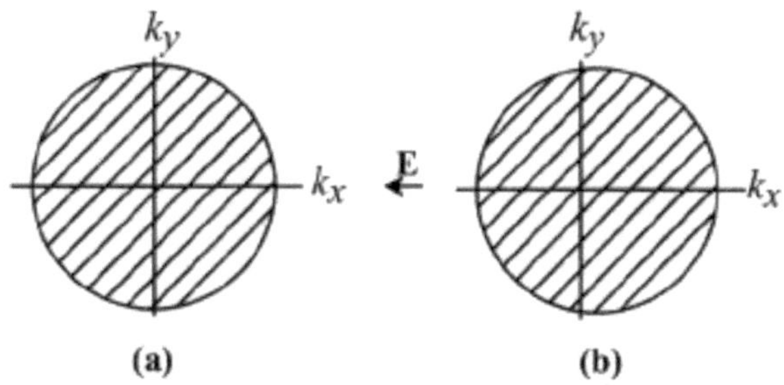

Fig. 4. Due to the electric field $\boldsymbol{E}$ pointed in the negative x-direction, the steady-state electron distribution in (b) is generated, which is a translation of the equilibrium distribution in (a) by the amount $\hbar^{-1}eE\tau$.

displaced, the conductivity σ depends on the particle density n. The Seebeck current is caused by the density difference in the thermally excited electrons near the Fermi surface, and hence, the thermal diffusion coefficient A depends on the density of states at the Fermi energy $\mathcal{N}_0$ [see Eq. (2.9)]. We further note that the diffusion coefficient D does not depend on m^* directly [see Eq. (2.7)]. Thus, the Ohmic and Seebeck currents are fundamentally different in nature.
For a single-carrier metal such as alkali metal (Na) which forms a body-centered-cubic (bcc) lattice, where only "electrons" exist, both $R_{\rm H}$ and S are negative. The *Einstein relation* between the conductivity σ and the diffusion coefficient D holds:

$$\sigma \propto D. \tag{2.12}$$

Using Eqs. (2.1) and (2.7), we obtain

$$\frac{D}{\sigma} = \frac{v_{\rm F}^2\tau/3}{q^2 n\tau/m^*} = \frac{2}{3}\frac{\varepsilon_{\rm F}}{q^2 n}, \tag{2.13}$$

which is a material constant. The Einstein relation is valid for a single-carrier system.

3. Applications

We consider two-carrier metals (noble metals). Noble metals including copper (Cu), silver (Ag) and gold (Au) form face-centered cubic (fcc) lattices. Each metal contains "electrons" and "holes". The Seebeck coefficient S for these metals are shown in Fig. 1. The S is positive for all

$$S > 0 \qquad \text{for Cu, Al, Ag}, \tag{3.1}$$

indicating that the majority carriers are "holes". The Hall coefficient R_H is known to be negative

$$R_H < 0 \quad \text{for Cu, Al, Ag}. \tag{3.2}$$

Clearly the Einstein relation (2.12) does not hold since the charge sign is different for S and R_H. This complication was explained by Fujita, Ho and Okamura (Fujita et al., 1989) based on the Fermi surfaces having "necks" (see Fig. 5). The curvatures along the axes of each

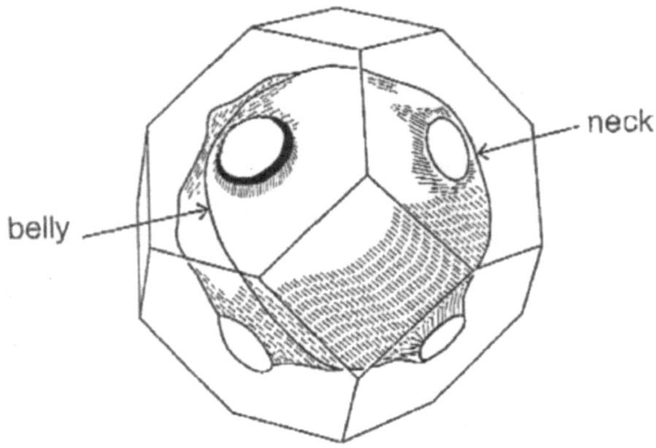

Fig. 5. The Fermi surface of silver (fcc) has "necks", with the axes in the $\langle 111 \rangle$ direction, located near the Brillouin boundary, reproduced after Ref. (Roaf, 1962; Schönberg, 1962; Schönberg & Gold, 1969).

neck are positive, and hence, the Fermi surface is "hole"-generating. Experiments (Roaf, 1962; Schönberg, 1962; Schönberg & Gold, 1969) indicate that the minimum neck area A_{111} (neck) in the k-space is 1/51 of the maximum belly area A_{111} (belly), meaning that the Fermi surface just touches the Brillouin boundary (Fig. 5 exaggerates the neck area). The density of "hole"-like states, n_{hole}, associated with the $\langle 111 \rangle$ necks, having the heavy-fermion character due to the rapidly varying surface with energy, is much greater than that of "electron"-like states, n_{electron}, associated with the $\langle 100 \rangle$ belly. The thermally excited "hole" density is higher than the "electron" density, yielding a positive S. The principal mass m_1 along the axis of a small neck ($m_1^{-1} = \partial^2 \varepsilon / \partial p_1^2$) is positive ("hole"-like) and large. The "hole" contribution to the conduction is small ($\sigma \propto m^{*-1}$), as is the "hole" contribution to Hall voltage. Then the "electrons" associated with the non-neck Fermi surface dominate and yield a negative Hall coefficient R_H.

The Einstein relation (2.12) does not hold in general for multi-carrier systems. The currents are additive. The ratio D/σ for a two-carrier system containing "electrons" (1) and "holes" (2) is given by

$$\frac{D}{\sigma} = \frac{(1/3)v_1^2\tau_1 + (1/3)v_2^2\tau_2}{q_1^2(n_1/m_1)\tau_1 + q_2^2(n_2/m_2)\tau_2}, \tag{3.3}$$

which is a complicated function of (m_1/m_2), (n_1/n_2), (v_1/v_2), and (τ_1/τ_2). In particular the mass ratio m_1/m_2 may vary significantly for a heavy fermion condition, which occurs whenever the Fermi surface just touches the Brillouin boundary. An experimental check on the violation of the Einstein relation can be be carried out by simply examining the T dependence of the ratio D/σ. This ratio D/σ depends on T since the generally T-dependent mean free times (τ_1, τ_2) arising from the electron-phonon scattering do not cancel out from

numerator and denominator. Conversely, if the Einstein relation holds for a metal, the spherical Fermi surface approximation with a single effective mass m^* is valid.
Formula (2.12) indicates that the thermal diffusion contribution to S is T-independent. The observed S in many metals is mildly T-dependent. For example, the coefficient S for Ag increases slightly before melting ($\sim$ 970 °C), while the coefficient S for Au is nearly constant and decreases, see Fig. 1. These behaviors arise from the incomplete compensation of the scattering effects. "Electrons" and "holes" that are generated from the complicated Fermi surfaces will have different effective masses and densities, and the resulting incomplete compensation of τ's (*i.e.*, the scattering effects) yields a T-dependence.

4. Graphene and carbon nanotubes

4.1 Introduction

Graphite and diamond are both made of carbons. They have different lattice structures and different properties. Diamond is brilliant and it is an insulator while graphite is black and is a good conductor. In 1991 Iijima (Iijima, 1991) discovered carbon nanotubes (graphite tubules) in the soot created in an electric discharge between two carbon electrodes. These nanotubes ranging 4 to 30 nanometers (nm) in diameter are found to have helical multi-walled structure as shown in Figs. 6 and 7 after the electron diffraction analysis. The tube length is about one micrometer (μm).

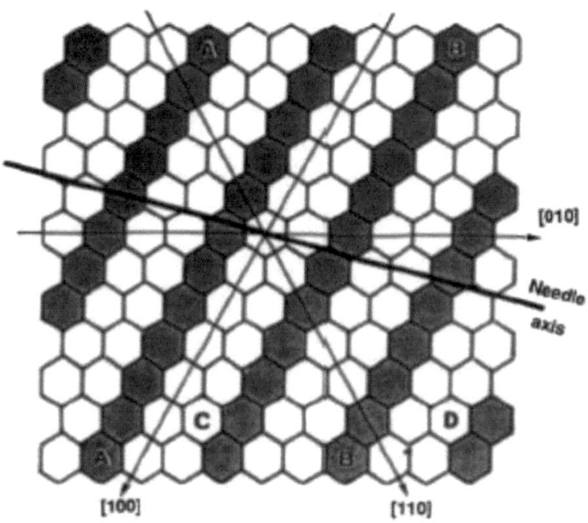

Fig. 6. Schematic diagram showing a helical arrangement of a carbon nanotube, unrolled (reproduced from Ref. (Iijima, 1991)). The tube axis is indicated by the heavy line and the hexagons labelled A and B, and A′ and B′, are superimposed to form the tube. The number of hexagons does not represent a real tube size.

The scroll-type tube shown in Fig. 7 is called the *multi-walled carbon nanotube* (MWNT). *Single-walled nanotube* (SWNT) shown in Fig. 8 was fabricated by Iijima and Ichihashi (Iijima & Ichihashi, 1993) and by Bethune *et al.* (Bethune et al., 1993). The tube size is about one nanometer in diameter and a few microns (μ) in length. The tube ends are closed as shown in Fig. 8. Unrolled carbon sheets are called *graphene*. They have honeycomb lattice structure as shown in Figs. 6 and 9. Carbon nanotubes are light since they are entirely made of light element carbon (C). They are strong and have excellent elasticity and flexibility. In fact, carbon fibers are used to make tennis rackets, for example. Today's semiconductor technology is based mainly on silicon (Si). It is said that carbon devices are expected to be as important or even more important in the future. To achieve this we must know the electrical transport properties of carbon nanotubes.

In 2003 Kang *et al.* (Kang et al., 2003) reported a logarithmic temperature (T) dependence of the Seebeck coefficient S in multiwalled carbon nanotubes at low temperatures ($T = 1.5$ K). Their data are reproduced in Fig. 10, where S/T is plotted on a logarithmic temperature scale after Ref. (Kang et al., 2003), Fig. 2. There are clear breaks in data around $T_0 = 20$ K. Above this temperature T_0, the Seebeck coefficient S is linear in temperature T:

$$S = aT\,, \qquad T > T_0 = 20\,\mathrm{K} \tag{4.1}$$

where $a = 0.15\ \mu\mathrm{V}/\mathrm{K}^2$. Below 20 K the temperature behavior is approximately

$$S \sim T \ln T, \qquad T < T_0\,. \tag{4.2}$$

The original authors (Kang et al., 2003) regarded the unusual behavior (4.2) as the intrinsic behavior of MWNT, arising from the combined effects of electron-electron interaction and

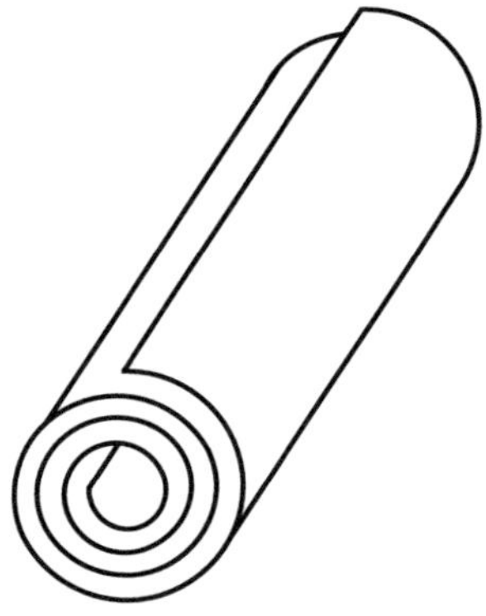

Fig. 7. A model of a scroll-type filament for a multi-walled nanotube.

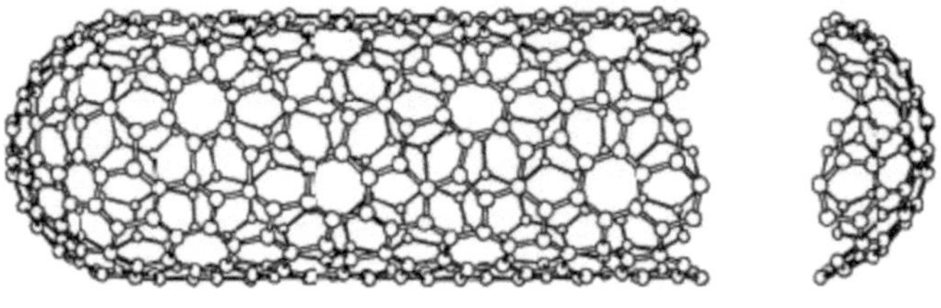

Fig. 8. Structure of a single-walled nanotube (SWNT) (reproduced from Ref. (Saito et al., 1992)). Carbon pentagons appear near the ends of the tube.

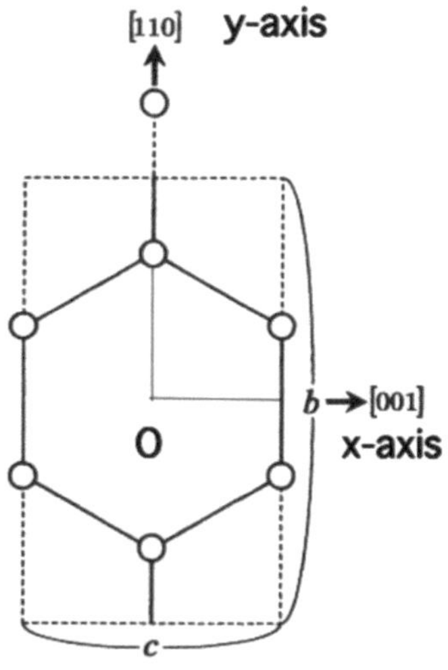

Fig. 9. A rectangular unit cell of graphene. The unit cell contains four C (open circle).

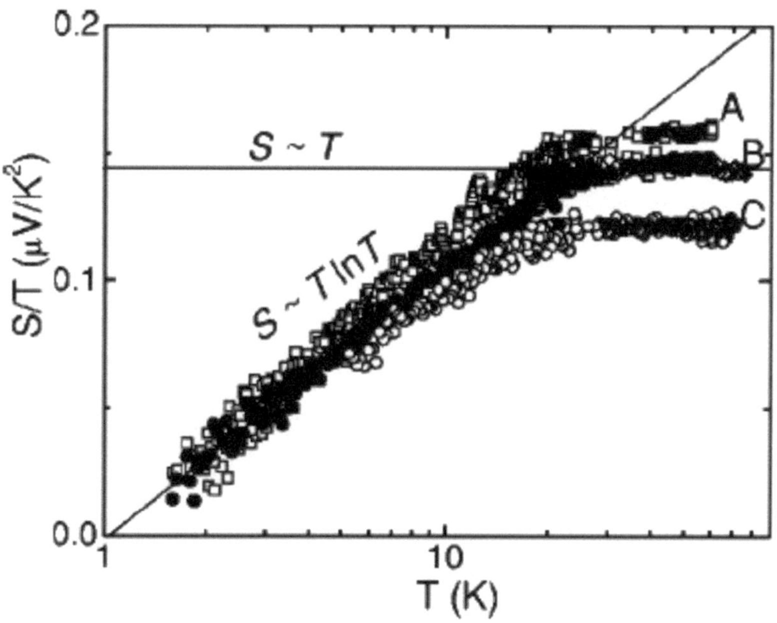

Fig. 10. A logarithmic temperature (T) dependence of the Seebeck coefficient S in MWNT after Ref. (Kang et al., 2003). A, B and C are three samples with different doping levels.

electron-disorder scattering. The effects are sometimes called as two-dimensional weak localization (2D WL) (Kane & Fisher, 1992; Langer et al., 1996). Their interpretation is based on the electron-carrier transport. We propose a different interpretation. Both (4.1) and (4.2) can be explained based on the Cooper-pairs (pairons) carrier transport. The pairons are generated by the phonon exchange attraction. We shall show that the pairons generate the T-linear behavior in (4.1) above the superconducting temperature T_0 *and* the $T \ln T$ behavior in (4.2) below T_0.

The current band theory of the honeycomb crystal based on the Wigner-Seitz (WS) cell model (Saito et al., 1998; Wigner & Seitz, 1933) predicts a gapless semiconductor for graphene, which is not experimentally observed. The WS model (Wigner & Seitz, 1933) was developed for the study of the ground-state energy of the crystal. To describe the Bloch electron motion in terms of the mass tensor (Ashcroft & Mermin, 1976) a new theory based on the Cartesian unit cell not matching with the natural triangular crystal axes is necessary. Only then, we can discuss the anisotropic mass tensor. Also phonon motion can be discussed, using Cartesian coordinate-systems, not with the triangular coordinate systems. The conduction electron moves as a wave packet formed by the Bloch waves as pointed out by Ashcroft and Mermin in their book (Ashcroft & Mermin, 1976). This picture is fully incorporated in our new theoretical model. We discuss the Fermi surface of graphene in section 4.2.

4.2 The Fermi surface of graphene

We consider a graphene which forms a two-dimensional (2D) honeycomb lattice. The normal carriers in the electrical charge transport are "electrons" and "holes." The "electron" ("hole") is a quasi-electron that has an energy higher (lower) than the Fermi energy *and* which circulates counterclockwise (clockwise) viewed from the tip of the applied magnetic field vector. "Electrons" ("holes") are excited on the positive (negative) side of the Fermi surface with the convention that the positive normal vector at the surface points in the energy-increasing direction.

We assume that the "electron" ("hole") wave packet has the charge $-e$ $(+e)$ and a size of a unit carbon hexagon, generated above (below) the Fermi energy ε_F. We will show that (a) the "electron" and "hole" have different charge distributions and different effective masses, (b) that the "electrons" and "holes" are thermally activated with different energy gaps $(\varepsilon_1, \varepsilon_2)$, and (c) that the "electrons" and "holes" move in different easy channels.

The positively-charged "hole" tends to stay away from positive ions C^+, and hence its charge is concentrated at the center of the hexagon. The negatively charged "electron" tends to stay close to the C^+ hexagon and its charge is concentrated near the C^+ hexagon. In our model, the "electron" and "hole" both have charge distributions, and they are not point particles. Hence, their masses m_1 and m_2 must be different from the gravitational mass $m = 9.11 \times 10^{-28}$ g. Because of the different internal charge distributions, the "electrons" and "holes" have the different effective masses m_1 and m_2. The "electron" may move easily with a smaller effective mass in the direction [110 c-axis]$\equiv$ [110] than perpendicular to it as we see presently. Here, we use the conventional Miller indices for the hexagonal lattice with omission of the c-axis index. For the description of the electron motion in terms of the mass tensor, it is necessary to introduce Cartesian coordinates, which do not necessarily match with the crystal's natural (triangular) axes. We may choose the rectangular unit cell with the side-length pair (b, c) as shown in Fig. 9. Then, the Brillouin zone boundary in the k space is unique: a rectangle with side lengths $(2\pi/b, 2\pi/c)$. The "electron" (wave packet) may move up or down in [110] to the neighboring hexagon sites passing over one C^+. The positively charged C^+ acts as a welcoming (favorable) potential valley center for the negatively charged "electron" while the same C^+ acts as a hindering potential hill for the positively charged "hole". The "hole" can however move easily over on a series of vacant sites, each surrounded by six C^+, without meeting the hindering potential hills. Then, the easy channel directions for the "electrons" and "holes" are [110] and [001], respectively.

Let us consider the system (graphene) at 0 K. If we put an electron in the crystal, then the electron should occupy the center O of the Brillouin zone, where the lowest energy lies. Additional electrons occupy points neighboring O in consideration of Pauli's exclusion principle. The electron distribution is lattice-periodic over the entire crystal in accordance with the Bloch theorem. The uppermost partially filled bands are important for the transport properties discussion. We consider such a band. The 2D Fermi surface which defines the boundary between the filled and unfilled k-space (area) is *not* a circle since the x-y symmetry is broken. The "electron" effective mass is smaller in the direction [110] than perpendicular to it. That is, the "electron" has two effective masses and it is intrinsically anisotropic. If the "electron" number is raised by the gate voltage, then the Fermi surface more quickly grows in the easy-axis (y) direction, say [110] than in the x-direction, i.e., [001]. The Fermi surface must approach the Brillouin boundary at right angles because of the inversion symmetry possessed by the honeycomb lattice. Then at a certain voltage, a "neck" Fermi surface must be developed.

The same easy channels in which the "electron" runs with a small mass, may be assumed for other hexagonal directions, [011] and [101]. The currents run in three channels $\langle 110 \rangle \equiv [110]$, [011], and [101]. The electric field component along a channel j is reduced by the directional cosine $\cos(\mu, j)$ $(= \cos\vartheta)$ between the field direction μ and the channel direction j. The current is reduced by the same factor in the Ohmic conduction. The total current is the sum of the channel currents. Then its component along the field direction is proportional to

$$\sum_{j\,\text{channel}} \cos^2(\mu, j) = \cos^2\vartheta + \cos^2(\vartheta + 2\pi/3) + \cos^2(\vartheta - 2\pi/3) = 3/2 \,. \tag{4.3}$$

There is no angle (ϑ) dependence. The current is isotropic. The number 3/2 represents the fact that the current density is higher by this factor for a honeycomb lattice than for the square lattice.

We have seen that the "electron" and "hole" have different internal charge distributions and they therefore have different effective masses. Which carriers are easier to be activated or excited? The "electron" is near the positive ions and the "hole" is farther away from the ions. Hence, the gain in the Coulomb interaction is greater for the "electron." That is, the "electron" are more easily activated (or excited). The "electron" move in the welcoming potential-well channels while the "hole" do not. This fact also leads to the smaller activation energy for the electrons. We may represent the activation energy difference by

$$\varepsilon_1 < \varepsilon_2 \,. \tag{4.4}$$

The thermally activated (or excited) electron densities are given by

$$n_j(T) = n_j e^{-\varepsilon_j / k_B T} \,, \tag{4.5}$$

where $j = 1$ and 2 represent the "electron" and "hole", respectively. The prefactor n_j is the density at the high temperature limit.

4.3 Single-walled nanotubes (SWNT)

Let us consider a long SWNT rolled with the graphene sheet. The charge may be transported by the channeling "electrons" and "holes" in the graphene wall. But the "holes" present inside the SWNT can also contribute to the charge transport. The carbon ions in the wall are positively charged. Hence, the positively charged "hole" can go through inside tube. In contrast, the negatively charged "electrons" are attracted by the carbon wall and cannot go straight in the tube. Because of this extra channel inside the carbon nanotube, "holes" can be the majority carriers in nanotubes although "electrons" are the dominant carriers in graphene. Moriyama *et al.* (Moriyama et al., 2004) observed the electrical transport in SWNT in the temperature range 2.6 - 200 K, and found from the field effect (gate voltage) study that the carriers are "holes".

The conductivity was found to depend on the pitch of the SWNT. The helical line is defined as the line in $\langle 100 \rangle$ passing the centers of the nearest neighbors of the C^+ hexagons. The helical angle φ is the angle between the helical line and the tube axis. The degree of helicity h may be defined as

$$h = \cos\varphi \,. \tag{4.6}$$

For a macroscopically large graphene the conductivity does not show any directional dependence (Fujita & Suzuki, 2010) as we saw in Sec. 4.2. The electrical conduction in SWNT depends on the pitch (Dai et al., 1996; Ebbesen et al., 1996) and can be classified into two groups: either semiconducting or metallic (Saito et al., 1998; Tans et al., 1997). This division in two groups arises as follows. A SWNT is likely to have an integral number of carbon hexagons around the circumference. If each pitch contains an integral number of hexagons, then the system is periodic along the tube axis, and "holes" (not "electrons") can move along the tube. Such a system is semiconducting and the electrical conduction is then characterized by an activation energy ε_2. The energy ε_2 has distribution since both the pitch and circumference have distributions. The pitch angle is not controlled in the fabrication processes. There are, then, more numerous cases where the pitch contains an irrational numbers of hexagons. In these cases the system shows a metallic behavior experimentally observed (Tans et al., 1998).

4.4 Multi-walled nanotubes (MWNT)

MWNT are open-ended. Hence, each pitch is likely to contain an irrational number of carbon hexagons. Then, the electrical conduction of MWNT is similar to that of metallic SWNT. The conductivity σ based on the pairon carrier model is calcullated as follows.

The pairons move in 2D with the linear dispersion relation (Fujita et al., 2009):

$$\varepsilon_{\rm p} = c^{(j)} p, \tag{4.7}$$

$$c^{(j)} = (2/\pi) v_{\rm F}^{(j)} \,, \tag{4.8}$$

where $v_{\rm F}^{(j)}$ is the Fermi velocity of the "electron" ($j = 1$) ["hole" ($j = 2$)].

Consider first "electron"-pairs. The velocity $\boldsymbol{v}$ is given by (omitting superscript)

$$\boldsymbol{v} = \frac{\partial \varepsilon_{\rm p}}{\partial \boldsymbol{p}} \quad \text{or} \quad v_x = \frac{\partial \varepsilon_{\rm p}}{\partial p}\frac{\partial p}{\partial p_x} = c\frac{p_x}{p} \,, \tag{4.9}$$

where we used Eq. (4.7) for the pairon energy $\varepsilon_{\rm p}$ and the 2D momentum,

$$p \equiv (p_x^2 + p_y^2)^{1/2} \,. \tag{4.10}$$

The equation of motion along the electric field E in the x-direction is

$$\frac{\partial p_x}{\partial t} = q'E, \tag{4.11}$$

where q' is the charge $\pm 2e$ of a pairon. The solution of Eq. (4.11) is given by

$$p_x = q'Et + p_x^{(0)}, \tag{4.12}$$

where $p_x^{(0)}$ is the initial momentum component. The current density $j_{\rm p}$ is calculated from (charge q') $\times$ (number density $n_{\rm p}$) $\times$ (average velocity $\bar{v}$). The average velocity $\bar{v}$ is calculated by using Eq. (4.9) and Eq. (4.12) with the assumption that the pair is accelerated only for the collision time τ *and* the initial-momentum-dependent terms are averaged out to zero. We then obtain

$$j_{\rm p} = q'n_{\rm p}\bar{v} = q'n_{\rm p}c\frac{\bar{p}_x}{p} = q'^2 n_{\rm p}\frac{c}{p}E\tau. \tag{4.13}$$

For stationary currents, the partial pairon density $n_{\rm p}$ is given by the Bose distribution function $f(\varepsilon_{\rm p})$:

$$n_{\rm p} = f(\varepsilon_{\rm p}) \equiv [\exp(\varepsilon_{\rm p}/k_{\rm B}T - \alpha) - 1]^{-1}, \tag{4.14}$$

where e^{α} is the fugacity. Integrating the current $j_{\rm p}$ over all 2D p-space, and using Ohm's law $j = \sigma E$, we obtain for the conductivity σ:

$$\sigma = (2\pi\hbar)^{-2} q'^2 c \int d^2p\, p^{-1} f(\varepsilon_{\rm p})\tau. \tag{4.15}$$

In the low temperatures we may assume the Boltzmann distribution function for $f(\varepsilon_{\rm p})$:

$$f(\varepsilon_{\rm p}) \simeq \exp(\alpha - \varepsilon_{\rm p}/k_{\rm B}T). \tag{4.16}$$

We assume that the relaxation time arises from the phonon scattering so that

$$\tau = (aT)^{-1}, \quad a = \text{constant}. \tag{4.17}$$

After performing the p-integration we obtain from Eq. (4.15)

$$\sigma = \frac{2}{\pi}\frac{e^2 k_{\rm B}}{a\hbar^2}e^{\alpha}, \tag{4.18}$$

which is temperature-independent. If there are "electrons" and "hole" pairons, they contribute additively to the conductivity. These pairons should undergo a Bose-Einstein condensation at lowest temperatures.

We are now ready to discuss the Seebeck coefficient S of MWNT. First, we will show that the S is proportional to the temperature T above the superconducting temperature T_0.

We start with the standard formula for the charge current density:

$$\boldsymbol{j} = q'n\bar{\boldsymbol{v}}, \tag{4.19}$$

where $\bar{\boldsymbol{v}}$ is the average velocity, which is a function of temperature T and the particle density n:

$$\bar{\boldsymbol{v}} = \boldsymbol{v}(n, T). \tag{4.20}$$

We assume a steady state in which the temperature T varies only in the x-direction while the density is kept constant. The temperature gradient $\partial T/\partial x$ generates a current:

$$j = q'n\frac{\partial v(n,T)}{\partial T}\frac{\partial T}{\partial x}\Delta x\,. \tag{4.21}$$

The thermal diffusion occurs locally. We may choose Δx to be a mean free path:

$$\Delta x = l = v\tau\,. \tag{4.22}$$

The current coming from the 2D pairon momentum $\boldsymbol{p}$, which is generated by the temperature gradient $\partial T/\partial x$, is thus given by

$$j_{\mathrm{p}} = q'n_{\mathrm{p}}\bar{v}_x(n_{\mathrm{p}},T) = q'n_{\mathrm{p}}\frac{\partial v}{\partial T}\frac{\partial T}{\partial x}v\tau\,. \tag{4.23}$$

Integrating Eq. (4.23) over all 2D p-space and comparing with Eq. (1.1), we obtain

$$\begin{aligned} A &= (2\pi\hbar)^{-2}q'\frac{\partial v}{\partial T}\int \mathrm{d}^2 p v_x f(\varepsilon_{\mathrm{p}})\tau \\ &= (2\pi\hbar)^{-2}q'\frac{\partial v}{\partial T}c\int \mathrm{d}^2 p \frac{p_x}{p} f(\varepsilon_{\mathrm{p}})\tau\,. \end{aligned} \tag{4.24}$$

We compare this integral with the integral in Eq. (4.15). It has an extra factor in p and generates therefore an extra factor T when the Boltzmann distribution function is adopted for $f(\varepsilon_{\mathrm{p}})$. Thus, we obtain

$$S = \frac{A}{\sigma} \propto T\,. \tag{4.25}$$

We next consider the system below the superconducting temperature T_0. The supercurrents arising from the condensed pairons generate no thermal diffusion. But non-condensed pairons can be scattered by impurities and phonons, and contribute to a thermal diffusion. Because of the zero-temperature energy gap

$$\varepsilon_{\mathrm{g}} \equiv k_{\mathrm{B}}T_{\mathrm{g}} \tag{4.26}$$

generated by the supercondensate, the population of the non-condensed pairons is reduced by the Boltzmann-Arrhenius factor

$$\exp(-\varepsilon_{\mathrm{g}}/k_{\mathrm{B}}T) = \exp(-T_{\mathrm{g}}/T)\,. \tag{4.27}$$

This reduction applies only for the conductivity (and not for the diffusion). Hence we obtain the Seebeck coefficient:

$$\frac{A}{\sigma} \propto \frac{T}{\exp(-T_{\mathrm{g}}/T)} = T\exp(T_{\mathrm{g}}/T)\,. \tag{4.28}$$

In the experiment MWNT bundles containing hundreds of individual nanotubes are used. Both circumference and pitch have distributions. Hence, the effective energy gap temperature T_{g} has a distribution. We may then replace (Jang et al., 2004)

$$\exp(T_{\mathrm{g}}/T) \quad \text{by} \quad (T'_{\mathrm{g}}/T)^{1/3} \tag{4.29}$$

where T_g' is a temperature of the order T_g. We then obtain

$$\frac{A}{\sigma} \propto T(T_g'/T)^{1/3}. \tag{4.30}$$

In summary, by considering moving pairons we obtained the T-linear behavior of the Seebeck coefficient S above the superconducting temperature T_c and the $T \ln T$-behavior of S at the lowest temperatures. The energy gap ε_g vanishes at T_c. Hence, the temperature behaviors should be smooth and monotonic as observed in Fig. 10. This supports the present interpretation based on the superconducting phase transition. The doping changes the pairon density and the superconducting temperature. Hence the data for A, B and C in Fig. 10 are reasonable.
Based on the idea that different temperatures generate different carrier densities and the resulting carrier diffusion generates a thermal electromotive force (emf), we obtained a new formula for the Seebeck coefficient (thermopower) S:

$$S = \frac{2 \ln 2}{\mathrm{d}} \frac{1}{qn} \varepsilon_F k_B \frac{\mathcal{N}_0}{\mathbb{V}},$$

where k_B is the Boltzmann constant, d the dimension, q, n, ε_F, $\mathcal{N}_0$ and $\mathbb{V}$ are charge, carrier density, Fermi energy, density of states at ε_F, and volume, respectively. Ohmic and Seebeck currents are fundamentally different in nature, and hence, cause significantly different transport behaviors. For example, the Seebeck coefficient S in copper (Cu) is positive, while the Hall coefficient is negative. In general, the Einstein relation between the conductivity and the diffusion coefficient does not hold for a multicarrier metal. Multi-walled carbon nanotubes are superconductors. The Seebeck coefficient S is shown to be proportional to the temperature T above the superconducting temperature T_0 based on the model of Cooper pairs as carriers. The S below T_0 follows a temperature behavior, $S/T \propto (T_g'/T)^{1/3}$, where $T_g' =$ constant, at the lowest temperatures.

5. Appendix: Derivation of Eq. (1.4)

In order to clearly understand diffusion let us look at the following simple situation. Imagine that four particles are in space a, and two particles are in space b as shown in Fig. 11. Assuming that both spaces a and b have the same volume, we may say that the particle density is higher in a than in b. We assume that half of the particles in each space will be heading toward the boundary CC′. It is then natural to expect that in due time two particles would cross the boundary CC′ from a to b, and one particle from b to a. This means that more particles would pass the boundary from a to b, that is, from the side of high density to that of low density. This is, in fact, the cause of diffusion.
The essential points in the above arguments are the reasonable assumptions that

(a) the particles flow out from a given space in all directions with the *same* probability, and

(b) the rate of this outflow is proportional to the number of particles contained in that space.

In the present case the condition (a) will be assured by the fact that each electron *collides* with impurities frequently so that it may lose the memory of how it entered the space originally and may leave with *no* preferred direction. In a more quantitative study it is found that the particle current $\boldsymbol{j}$ is proportional to the density gradient $\boldsymbol{\nabla} n$:

$$\boldsymbol{j} = -D\boldsymbol{\nabla} n, \tag{A.1}$$

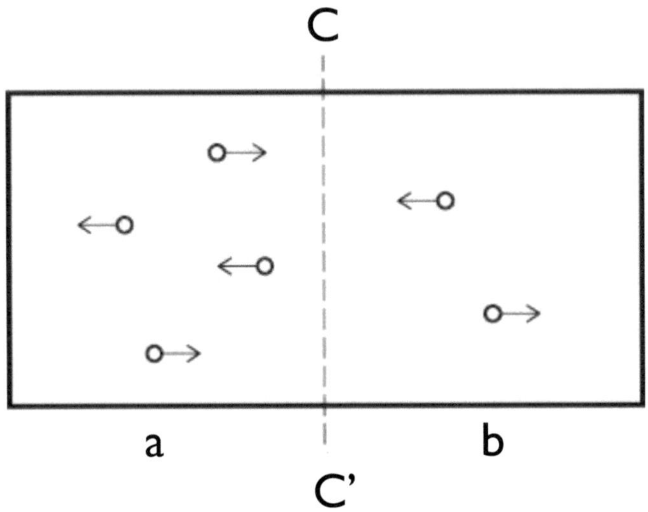

Fig. 11. If the particles flow out in all directions with no preference, there will be more particles crossing the imaginary boundary CC′ in the a to b direction than in the opposite direction.

where D is the diffusion coefficient. This linear relation (A.1) is called *Fick's law*.
Consider next *thermal conduction*. Assume that the spaces a and b are occupied by the same numbers of the particles. Further assume that the temperature T is higher in b than in a. Then, the particle speed is higher in b than in a in the average. In due time a particle crosses the boundary CC′ from a to b and another crosses the boundary CC′ from b to a. Then, the energy is transferred through the boundary. In a more detailed study *Fourier's law* is observed:

$$\boldsymbol{q} = -K\boldsymbol{\nabla}T, \tag{A.2}$$

where $\boldsymbol{q}$ is the heat (energy) current and K is called the *thermal conductivity*.
We now take a system of free electrons with mass m and charge $-e$ immersed in a uniform distribution of impurities which act as scatterers. We assume that a free classical electron system in equilibrium is characterized by the ideal gas condition so that the average electron energy ε depends on the temperature T only:

$$\varepsilon = \varepsilon(n, T) = \varepsilon(T), \tag{A.3}$$

where n is the electron density. The electric current density $\boldsymbol{j}$ is given by

$$\boldsymbol{j} = (-e)n\boldsymbol{v}, \tag{A.4}$$

where $\boldsymbol{v}$ is the velocity field (average velocity). We assume that the density n is constant in space and time. If there is a temperature gradient, then there will be a current as shown below. We assume first a one-dimensional (1D) motion. The velocity field v depends on the temperature T, which varies in space.
Assume that the temperature T is higher at $x + \Delta x$ than at x:

$$T(x + \Delta x) > T(x). \tag{A.5}$$

Then

$$v[n,T(x+\Delta x)] - v[n,T(x)] = \frac{\partial v(n,T)}{\partial T}\frac{\partial T}{\partial x}\Delta x\,. \tag{A.6}$$

The diffusion and heat conduction occur locally. We may choose Δx to be a mean free path

$$l = v\tau\,, \tag{A.7}$$

which is constant in our system. Then the current j is, from Eq. (A.4),

$$j = (-e)n\frac{\partial v}{\partial T}l\frac{\partial T}{\partial x}\,. \tag{A.8}$$

Using Eqs. (1.1), (A.7) and (A.8), we obtain

$$A = (-e)n\frac{\partial v}{\partial T}v\tau\,. \tag{A.9}$$

The conductivity σ is given by the Drude formula:

$$\sigma = e^2\frac{n}{m}\tau\,. \tag{A.10}$$

Thus, the Seebeck coefficient S is, using Eqs. (A.9) and (A.10),

$$\begin{aligned} S = \frac{A}{\sigma} &= -\frac{1}{ne}m\frac{\partial v}{\partial T}\frac{l}{\tau} = -\frac{1}{ne}m\frac{\partial v^2}{\partial T} \\ &= -\frac{1}{ne}\frac{\partial}{\partial T}\left(\frac{1}{2}mv^2\right) = -\frac{1}{ne}\frac{\partial \varepsilon}{\partial T} = -\frac{1}{ne}c\,, \end{aligned} \tag{A.11}$$

where

$$c \equiv \frac{\partial \varepsilon}{\partial T}\,. \tag{A.12}$$

is the heat capacity per electron.

Our theory can simply be extended to a 3D motion. The equipartition theorem holds for the classical electrons:

$$\left\langle \frac{1}{2}mv_x^2 \right\rangle = \left\langle \frac{1}{2}mv_y^2 \right\rangle = \left\langle \frac{1}{2}mv_z^2 \right\rangle = \frac{1}{2}k_{\mathrm{B}}T\,, \tag{A.13}$$

where the angular brackets mean the equilibrium average. Hence the average energy is

$$\varepsilon \equiv \frac{1}{2}mv^2 = \frac{1}{2}(v_x^2 + v_y^2 + v_z^2) = \frac{3}{2}k_{\mathrm{B}}T\,. \tag{A.14}$$

We obtain

$$A = -en\frac{1}{2}\frac{\partial v^2}{\partial t}\tau\,. \tag{A.15}$$

Using this, we obtain the Seebeck coefficient for a 3D motion as

$$S = \frac{A}{\sigma} = -\frac{c_{\mathbb{V}}}{3ne} = -\frac{k_{\mathrm{B}}}{2e}\,, \tag{A.16}$$

where

$$c_{\mathbb{V}} \equiv \frac{\partial \varepsilon}{\partial T} = \frac{3}{2}k_{\mathrm{B}} \tag{A.17}$$

is the heat capacity per electron. The heat capacity per unit volume, $c_{\mathbb{V}}$, is related by the heat capacity per electron, c, by

$$c_{\mathbb{V}} = nc\,. \tag{A.18}$$

6. References

Ashcroft, N.W. & Mermin, N.D. (1976). *Solid State Physics* (Saunders, Philadelphia), pp. 256–258, 290–293.

Bethune, D.S., Kiang, C.H., de Vries, M.S., Gorman, G., Savoy, R., Vazquez, J. & Beyers, R. (1993). Cobalt-catalysed growth of carbon nanotubes with single-atomic-layer walls, *Nature* Vol. 363, 605–607.

Dai. H., Wong, E.W. & Lieber, C.M. (1996). Probing Electrical Transport in Nanomaterials: Conductivity of individual Carbon Nanotubes, *Science* Vol. 272, 523–526.

Ebbesen, T.W., Lezec. H.J., Hiura, H., Bennett, J.W., Ghaemi, L.J. & Thio, T. (1996). Electrical conductivity of individual carbon nanotubes, *Nature* Vol. 382, 54–56.

Fujita, S., Ho, H-C. & Okamura, Y. (2000). Quantum Theory of the Seebeck Coefficient in Metals, *Int. J. Mod. Phys.* B Vol. 14, 2231–2240.

Fujita, S., Ito, K. & Godoy, S. (2009). *Quantum Theory of Conducting Matter. Superconductivity* (Springer, New York) pp. 77–79.

Fujita, S. & Suzuki, A. (2010). Theory of temperature dependence of the conductivity in carbon nanotubes, *J. Appl. Phys.* Vol. 107, 013711–4.

Iijima, S. (1991). Helical microtubules of graphitic carbon, *Nature* Vol. 354. 56–58.

Iijima, S. & Ichihashi, T. (1993). Single-shell carbon nanotubes of 1-nm diameter, *Nature* Vol. 363, 603–605.

Jang, W.Y., Kulkami, N.N., Shih, C.K. & Yao, Z. (2004). Electrical characterization of individual carbon nanotubes grown in nano porous anodic alumina templates, *Appl. Phys. Lett.* Vol. 84, 1177–1180.

Kane, C. L. & Fisher, M. P. A. (1992). Transport in a one-channel Luttinger liquid, *Phys. Rev. Lett.* Vol. 68, 1220–1223.

Kang, N, Lu, L., Kong, W.J., Hu, J.S., Yi, W., Wang, Y.P., Zhang, D.L., Pan, Z.W & Xie, S.S. (2003). Observation of a logarithmic temperature dependence of thermoelectric power in multi wall carbon nanotubes, *Phys. Rev.* B Vol. 67, 033404–4.

Langer, L., *et al.* (1996). Quantum Transport in a Multiwalled Carbon Nanotube, *Phys. Rev. Lett.* Vol. 76, 479–482.

Moriyama, S., Toratani, K., Tsuya, D., Suzuki, M. Aoyagi, Y. & Ishibashi, K. (2004). Electrical transport in semiconducting carbon nanotubes, *Physica* E Vol. 24, 46–49.

Roaf, D.J. (1962). The Fermi Surface of Copper, Silver and Gold II. Calculation of the Fermi Surfaces, *Phil. Trans. R. Soc. Lond.* Vol. 255, 135–152.

Rossiter, P.L. & Bass, J. (1994). *Metals and Alloys.* in *Encyclopedia of Applied Physics* 10, (Wiley-VCH Publ., Berlin), pp. 163–197.

Saito, R., Fujita, M., Dresselhaus, G. & Dresselhaus, M.S. (1992). Electronic structure of chiral graphene tubles, *Appl. Phys. Lett.* Vol. 60, 2204–2206.

Saito, R.; Dresselhaus, G. & Dresselhaus, M.S. (1998). *Physical Properties of Carbon Nanotubes* (Imperial College, London) pp. 156–157.

Schönberg, D. (1962). The Fermi Surfaces of Copper, Silver and Gold I. The de Haas-van Alphen Effect, *Phil. Trans. R. Soc. Lond.* Vol. 255, 85–133.

Schönberg, D. & Gold, A.V. (1969). *Physics of Metals-1*, in *Electrons*, ed. Ziman, J.M. (Cambridge University Press, UK), p. 112.

Tans, S.J., Devoret, M.H., Dai, H., Thess, A., Smalley, R., Geerligs, L.J. & Dekker. C. (1997). Individual single-wall carbon nanotubes as quantum wires, *Nature* Vol. 386, 474–477.

Tans, S.J., Vershueren, A.R.M. & Dekker, C. (1998). Room-temperature transistor based on a single carbon nanotube, *Nature* Vol. 393, 49–52.

Wigner, E. & Seitz, F. (1933), On the Constitution of Metallic Sodium, *Phys. Rev.* Vol. 43, 804–810.

2

Electromotive Forces in Solar Energy and Photocatalysis (Photo Electromotive Forces)

A.V. Vinogradov[1,2], V.V. Vinogradov[2], A.V. Agafonov[1,2],
A.V. Balmasov[3] and L.N. Inasaridze[3]
[1]*Department of Ceramic Technology and Nanomaterials, ISUCT,*
[2]*Laboratory of Supramolecular Chemistry and Nanochemistry, SCI RAS,*
[3]*Department of Electrochemistry ISUCT,*
Russia

1. Introduction

The photoelectric polarization method is based on the inner photoeffect phenomenon which can be observed upon illumination of a photoactive material. Upon illumination of an oxide in its own region of optical absorption the arising non-equilibrium electrons and holes can be spatially separated within the surface oxide phase in a way when on one of interface boundaries there appears an excess of nonequilibrium negative charges, and on the other – an excess of positive charges. The photoelectric polarization emf arising as a result of charge carrier separation can be measured. Thus, the inner photoeffect is a structure-sensitive property of compounds. The inner photoelectric effect is of interest, on the one hand, as a factor that is responsible for a number of electrochemical and corrosion effects arising upon the exposure of metal and semiconductor electrodes to irradiation. On the other hand, it can be used for obtaining information on the nature and character of processes proceeding on the real materials. Thus, this method can be widely used for the evaluation of photoactivity of modern solar elements and photochemical converters of solar energy. The pathways for charge collection are much shorter, allowing the use of inexpensive low-quality materials, and also of organic semiconductors in which light absorption generates not free charge carriers but short-lived excitons that must reach an interface in order to separate at it and generate photocurrent and photo-emf. Thus, in this chapter we will consider the principles and peculiarities of the arising of the photo-emf in porous nanoarranged coatings using the most practiced synthesis methods: sol-gel method, polymer-assisted synthesis and electrochemical precipitation. At the same time, photocatalysis is closely related to photoelectrochemistry, and the fundamentals of both disciplines are covered in this volume, as among the key objects described there have been chosen the films on the basis of nanostructured titania that is widely used both as a catalyst and a component of solar elements. Finally, we will describe the measurement of electron-transfer dynamics at the molecule/semiconductor interface, and cover techniques for the characterization of photoelectrochemical titania-based systems.

2. Fundamentals in photoelectrochemistry

Titania-based preparations occupy nowadays leading positions both in the field of industrial photocatalyst manufacturing (Hombikat, Degussa P-25, P-90, etc.), and in the sphere of scientific studying. The basic direction of researches is the determination of approaches to increasing the photoactivity [Vinogradov et al., 2008, 2009, 2010]. Thus quantization effects play decisive role in the processes of generating the electron-hole pairs. Nanosized TiO_2 particles are of outstanding importance in this context. When electrons and holes are confined by potential barriers to small regions of space where the dimensions of the confinement are less than the de Broglie wavelength of these charge carriers, pronounced quantization effects develop; the length scale below which strong quantization effects begin to occur ranges from about 5 nm to 25 nm for typical semiconductors.

Among the most widespread methods of obtaining the colloidal semi-conductor nanoparticles the sol-gel technology occupies leading position. In the papers by Agafonov et al., there was shown a manifestation for high photoactivity of nanodisperse TiO_2 particles obtained by titanium isopropylate hydrolysis, which was estimated using data of photopolarization measurements. The use of the given technique allowed to achieve both optimum parameters for comparison of photoactivity of synthesized preparations, and the deepest interpretation of studied properties.

The main factor that determines unique photoactivity properties of titania-based materials is the dispersion of used preparations. Using semiconductor particles in the process of photocatalytic reactions is possible only in the case of the presence of a highly developed surface, under conditions of separating the formed electron-hole pairs without recombination at their movement from bulk to surface. Besides, the more developed a surface is, the more difficult it is for carriers to unite again. At the same time oxidation and reduction reactions should take place simultaneously on a particle surface (otherwise the particle will be charged, and the reaction will stop). The limiting stage will therefore be the rate of chemical reaction. Thus, the particle acts as a microelectrode, keeping the potential of anodic and cathodic electric currents that are equal in magnitude. When using large semiconductor particles, currents formed in them have insignificant magnitude in the darkness under open chain conditions as the basic density of charge carriers (for example, electrons in an n-type semiconductor) on a surface will be minimal because of long distances of movement, as is shown in $d << d_{sc}$, where d_{sc} is a thickness of charge transfer area. At the same time, in the case of very small sizes of particles there takes place a reverse procedure, because there is not enough room for formation of charges in the bulk, $d << d_{sc}$. After a slight light excitation, insignificant charge carriers (for example, holes in n-type semiconductors) in the largest particles become electron donors in solutions, and it leads to a negative charge of the particle and provides a positive charge of the entire complex. Thus, the combination of these two processes leads to the mutual leveling in the energy of the entire system.

In a semiconductor with small particles ($d << d_{sc}$), the photogenerated electrons and holes can easily move to the surface and react with electrons and holes of acceptors, provided that energetic leveling is observed.

3. New inorganic materials – perspective for solar energy conversion

While science development stimulated essential interest in the field of photo- and electrochemistry, considerable progresses in the increase in sensitivity and depreciation of

solar elements on the basis of solid-state photogalvanic cells have been made. Thus, the understanding of such a progress can be reached if we consider the basic fundamental concepts. Using solid-state cells demands direct contact between two phases of substances with different mechanisms of conductivity. Metal–semiconductor contact can be provided by a Schottky barrier while semi-conductor layers with different polarity of carriers provide p-n type. Excitation of an electron-hole pair as a result of a photon absorption by the semiconductor is possible in such systems if an energy of a photon is more than the bandgap ($h\lambda > E_g$). In this case, charge carriers at the interface can be separated effectively into separate electrons and holes, and that in turn increases the currents in the external contour. In such materials the conductivity of solid-state particles is electronic as a rule. Intensive researches during the last two decades have led to the inevitable conclusion that a rather narrow bandgap promoting phototransformation of visible light is peculiar for preparations with weaker chemical bond in the semiconductor, and that leads to the processes of self-oxidation and photocorrosion, which destroys used materials. The solution of this problem is probable by monitoring the separation of light absorption and charge separation functions, by sub-bandgap sensitization of the semiconductor with an electroactive dye. A wide bandgap is peculiar for a stable semiconductor, such as titanium dioxide with E_g = 3.1 eV, which therefore normally exhibits a photovoltaic response only under ultraviolet irradiation, can then photorespond to visible light of wavelength 400 - 750 nm, or 1.6 - 3.0 eV photons.
Impurity-induced conductance changes are therefore often much smaller than expected. In fact, in many 'doped' nanoporous films, the observed conductivity is found to be due to a hopping-type defect conduction mechanism, and may therefore be of only limited use in devices. The top-view image (Figure 1a) is coherent with disordered crystalline nanoparticles with narrow particle size distribution, approximately 10 nm. According to the general diffraction data (Figure 1b), the material is constructed from the anatase-brookite form crystallites, with size of about 5 nm (according to ring broadening). According to the low-temperature nitrogen adsorption - desorption data (Nova 1200e), the specific surface area of such a material amounts to 162 m^2/g, fig. 1c.

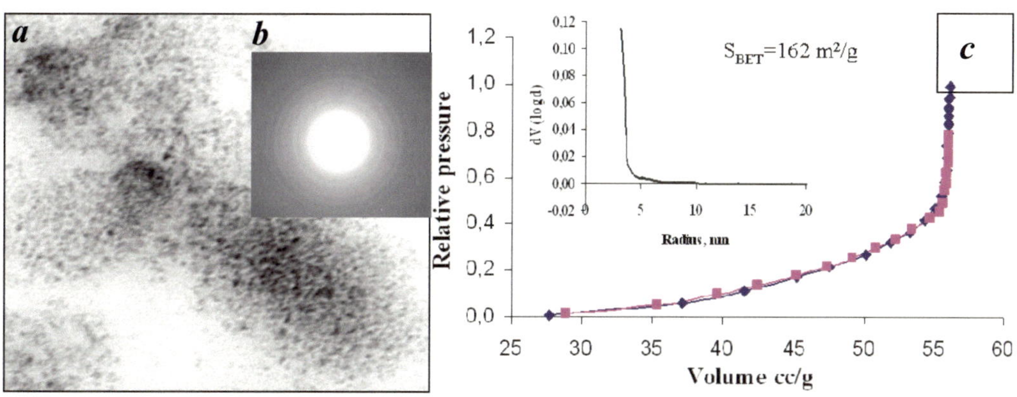

Fig. 1. The TEM images of titania film without silver nanoparticles: a) top-view; b) electron transmission diffraction pattern; c) adsorption – desorption isotherms of nitrogen and pore size distribution.

Figure 2a shows AFM micrographs of porous nanocrystalline anatase TiO_2 films with a grain size of approximately 10 nm. The volume fraction in these films is about 50% and measurements by the BET method show that the internal surface area is several hundred times the planar area for a 5 thick film.

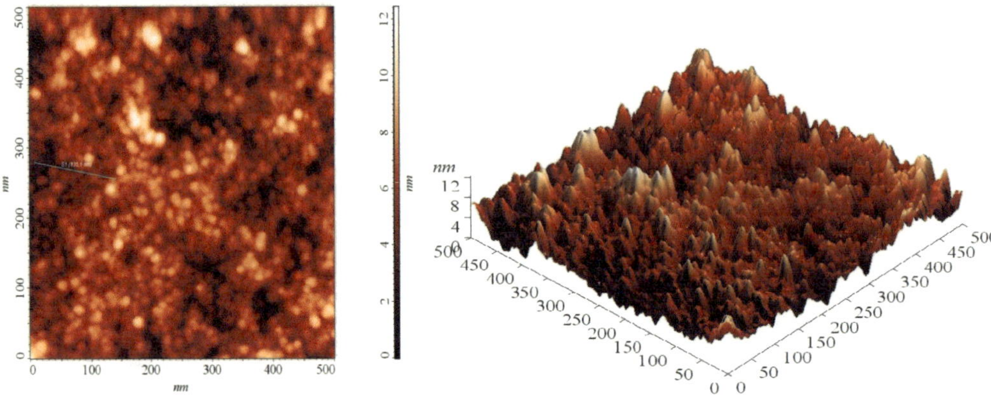

(a) (b)

Fig. 2. (a) The AFM micrograph of nanoporous TiO_2 film formed by spherical nanoparticles with narrow size distribution.

4. Method of photoelectric polarization

The essence of the photoelectric effect is as follows: when light of the corresponding wavelength and energy is absorbed by a crystal, from its surface electrons are emitted. Action of usual photocells is based on this principle. If a material is in vacuum, then it appears possible to collect emitted electrons, applying certain voltage. The resulting current force is the measure for quantity of absorbed light.

In the second half of the 20th century Russian scientists E.K.Osche and I.L.Rosenfeld suggested using the method of measuring the photoelectric polarization for determining the kinetics of electrode reactions occurring upon anodic oxidation and metal passivation, and also for estimating the structural and semiconductor properties of metals. Oxides on the surface of metals are compounds of variable composition for which the deviation from stoichiometry is the main and natural property. Thus, depending on character of such a deviation, i.e. on whether excess metal or oxygen prevails in the lattice, an oxide can possess electronic or hole type of conductivity. Degree of the deviation from stoichiometry, i.e. how much concentration of one of the excess components exceeds another, determines the concentration of free charge carriers in an oxide [Osche et al., 1969].

The internal photoelectric effect is of interest, on the one hand, as the factor responsible for a number of electrochemical and corrosion effects arising upon irradiation of metal and semiconductor electrodes. On the other hand, it can be used in structurally-sensitive photoelectric methods for obtaining information on the nature and character of the processes occurring on the real electrodes [Osche et al.,1978].

5. Technique of measuring the photoelectric polarization

The PEP method is based on the phenomenon of internal photoeffect observed upon illumination of an electrode placed in an electrolyte. Under the influence of light, in the surface layer there arise electron-hole pairs which are spatially separated in the electric field of impoverished layer: the electrons move deep into the semiconductor, and the holes close to the surface, reducing the magnitude of the surface charge. The bulk spatial charge is formed, therefore from the direction of irradiated contact the Schottky barrier magnitude decreases, and the height of the second barrier does not change. Simultaneously the electrons grabbed by the adsorbed oxygen atoms on the surface are released and move towards the conductivity zone, and the holes move to the valence zone, thereby reducing barriers on boundaries between particles. Because of the decrease in barrier, on the electrodes there arises a potential difference that is equal to the observed photo-emf, and the electrons from near-contact areas tunnel into the semiconductor, thereby generating a photocurrent [Vakalov et al. 2010].

The block diagram of installation for measuring the photo-emf is shown in Fig. 3. The photo-emf measurements are carried out in the usual electrochemical cell 1, in which except the electrode under study 2 the auxiliary electrode of the platinized platinum 3 is placed. Illumination of an investigated electrode is performed by rectangular impulses of non-spread light of a mercury lamp 7 through the quartz lens 6 and the quartz glass 4. The quartz lamp is turned on using the incendiary device 9. Duration of a light impulse is set by means of the photoshutter 5 and amounts to $5 \cdot 10^{-3}$ s. For the registration of the photo-emf arising in the surface layer upon pulse illumination, there serves the oscillograph 11, on the screen of which the sign and amplitude of the photoresponse are observed. The photoelectric signals from the cell are amplified using the amplifier 10. Wires in the alternating voltage chain should be contained in a metal braid and have the minimum length. While these conditions are observed allowing to reduce the level of the extraneous noise to a minimum, the registering scheme provides sensitivity to $5 \cdot 10^{-6}$ V.

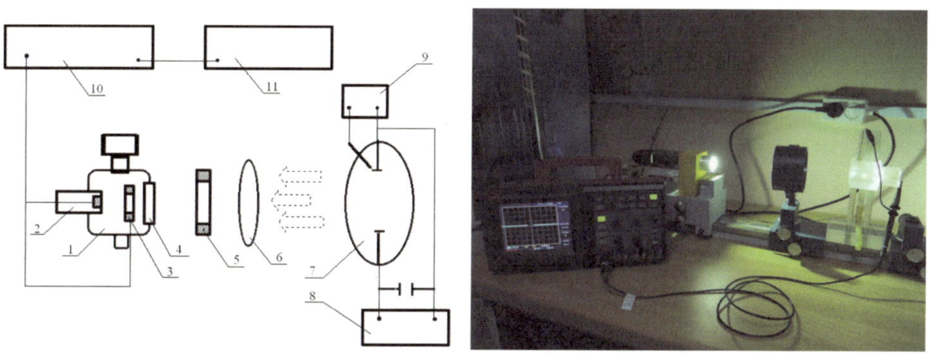

Fig. 3. The scheme of installation for measuring the photoelectric polarization: 1 - cell; 2 - working electrode; 3 - auxiliary electrode; 4 - quartz glass; 5 - photoshutter; 6 - quartz lens; 7 - DRS-250 mercury lamp; 8 - VSA power source; 9 - lamp incendiary device; 10 - amplifier UC-28, 11 - oscillograph S1-69.

6. Interpretation of data on the photoresponse arising in nanomaterials

The internal photoeffect belongs to structurally sensitive properties of a crystal. Therefore there is a basic possibility of using the internal photoeffect for obtaining the information on defective structure of an oxide, in particular, on the character and degree of deviation from stoichiometry. The surface oxide on metals is as a rule accepted to have the constant

composition corresponding to the stoichiometric formula of compounds. Meanwhile, metal oxides are compounds of variable composition, for which the deviation from stoichiometry is thermodynamically caused phenomenon. Depending on the surrounding conditions (pressure of oxygen, temperature) such compounds are capable to change the ratio of excess metal and oxygen in their crystal lattices within the considerable bounds without formation of a new phase. So, for example, the titania phase, whose deviation from stoichiometry is caused by the loss of balance of anionic and cationic vacancies, remains stable in the range of structures $TiO_{1.35}$–$TiO_{0.69}$. The other oxides suppose much less deviations from stoichiometry without formation of a new phase. Such compounds, depending on the character of deviation from stoichiometry, can possess n- or p- conductivity type. Degree of deviation from stoichiometry determines the concentration of own nuclear defects and of free charge carriers in an oxide, charge and substance transport and reactivity of an oxide. The most insignificant deviations from stoichiometry lead to a sharp change in physical and chemical properties of an oxide. For example, electrical conductivity of stoichiometric oxide TiO_2 is 10^{-10} Om^{-1} cm^{-1}, and that of non-stoichiometric one $TiO_{1.9995}$ is 10^{-1} Om^{-1} cm^{-1}.
The calculation of the metal surface oxide composition degree of deviation from stoichiometry on the basis of measurements of photoelectric polarization is performed in [Osche et al., 1978]. For the calculation of the stationary electromotive force of photoelectric polarization let us write down the concentrations of darkening electrons and holes as follows:

$$p_0 = N_B \cdot \exp(E_B - F_0), \tag{1}$$

$$n_0 = N_C \cdot \exp[-(E_C - F_0)], \tag{2}$$

where N_B and N_C is the density of states in valence and free zones; E_B is the energy of an upper part of the valence zone; E_C is the energy of a bottom part of free zone; F_0 is a Fermi's level.
By analogy, for non-equilibrium holes and electrons it is possible to write down:

$$p = N_B \cdot \exp\left(E_B - F_p\right), \tag{3}$$

$$n = N_C \cdot \exp[-(E_C - F_n)], \tag{4}$$

where F_p and F_n is a hole and an electron Fermi quasilevel.
From the equation system (4, 5) we have:

$$\ln\frac{p_0}{n_0} = \ln\frac{N_B}{N_C} + (E_B - F_0) + (E_C - F_0).$$

By analogy, for the equation system (3, 4) we have:

$$\ln\frac{p}{n} = \ln\frac{N_B}{N_C} + \left(E_B - F_p\right) + (E_C - F_n).$$

Thus, the stationary EMF of photoelectric polarization arising upon illumination is equal to:

$$V_{PEP} = \ln\frac{p}{n} - \ln\frac{p_0}{n_0} = (F_n - F_0) - (F_0 - F_p) \quad (5)$$

or in Volts:

$$V_{PEP} = \frac{k \cdot T}{e} \cdot \left(\ln\frac{p}{n} - \ln\frac{p_0}{n_0} \right). \quad (6)$$

7. Photoresponse in TiO_2-based nanomaterials obtained using different methods

Photoelectrical properties of wide bandgap metal oxide (TiO_2, ZnO, etc.,) thin films have drawn a great deal of attention in recent years due to their wide application in solar cells and photocatalysts [Gratzel et al., 1991; Masakazu, 2000]. Titanium dioxide is one of the promising candidates in the dye-sensitized [Li et al., 1999], conjugated polymer [Kwong et al., 2004] and inorganic semiconductor [Rincon et al., 2001] based solar cell applications. Presently many research groups are involved in improving the photoconduction and photovoltaic efficiency of the TiO_2 thin films by enhancing the charge carrier transport and by reducing the recombination centers. Titanium dioxide exhibits polymorphs such as anatase, rutile and brookite. Among the above structures, anatase exhibits higher photoactivity than rutile and brookite. Usually as deposited TiO_2 thin films are amorphous and photoinactive in nature. To achieve the photoactivity in these films structural transformation from amorphous to anatase phase is necessary. Thermoannealing is one of the suitable post-treatments to attain the phase transformation from amorphous to crystalline structure. During the thermoannealing processes the oxidation state of 2pO valence bands is modified due to the energy contribution from anharmonic electron-phonon interaction [Kityk et al., 2001] and it leads to reduction of Ti^{4+} states to Ti^{3+} states. Moreover, the critical energy necessary for such process even given by IR induced principle corresponds to about 420°C [Kityk et al., 2005], which is confirmed in the present work by photo transient decay spectra of TiO_2 films annealed at 425°C. Creation of this oxygen vacancies (Ti^{3+}) act as a trap levels in TiO_2 layers and it influences the efficiency of the dye-sensitized solar cells [Weidmann et al., 1998]. The knowledge of the trap levels and study of their nature will lead to understand the efficiency limiting parameters in the solar cells. Thermally stimulated current (TSC) measurement is a well-known non-isothermal technique for the investigation of trap levels in semiconducting materials [Zeenath et al., 2000; Pai et al., 2007]. This permits to determine the gap states and their capture cross section. The study of photo transient decay provides an understanding of photogenerations and transport of free carriers in the solid.

However, recent publications on obtaining the photoactive titania of anatase-brookite crystal form from a solution by using temperature dehydration [Alphonse et al., 2010] have allowed to essentially expand the spectrum of using titania in combination with organic sensitizers and metal nanoparticles while creating solar cells. Thus, in this chapter of the monograph we will consider the basic approaches on the establishment of the reasons of photo-emf emergence in the thin TiO_2 films obtained using the most widely used and modern methods, such as template synthesis, sol-gel technology with ultrasonic treatment, anodic electrochemical precipitation, precipitation of the layered heterostructures containing metal nanoparticles or organic dyes.

8. Sol-gel technology employing ultrasonic treatment

As a basic method of sol-gel synthesis we have used an approach, in which stabilization of hydrolysis process was performed by regulation of pH with formation of colloidal nanocluster system, which can develop into gel (pH 2–6, formation of macroscopically oriented structures) or sol (pH > 6, nanosized metal-polymer complexes). Modifying was performed using sonochemical treatment. Sol formation took place upon thermal treatment at 80°C. Further calcination led to the formation of crystallized nanoparticles, see Fig. 4.

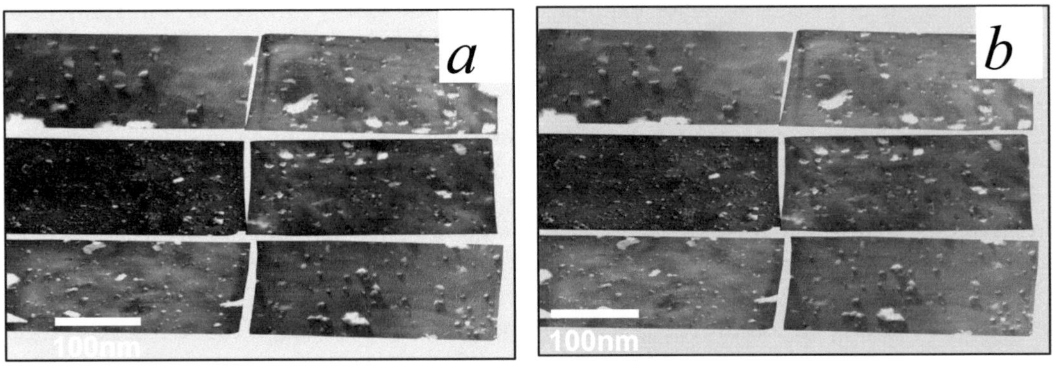

Fig. 4. The TEM images of titania powders, prepared a) with USI and diethylamine, b) with USI and acetic acid.

As investigations have revealed, using ultrasonic treatment, it is possible to substantially increase the photoactivity of synthesized preparations. The reason for this is correlation of the structure formed upon intensification of olation and oxolation processes, which in turn promotes obtaining highly stable sols that form defectless nanocrystals, as a rule, in anatase form. Absence of the stage of thermal treatment in the given synthesis method results, according to X-ray analysis data, in an amorphous phase. Calcination of films led to an increase in photo-emf by ten or more times, that is related to the formation of crystal phase. Using diethylamine as the initiator of hydrolysis as compared to ice acetic acid promotes sharp increase in the photoresponse indices and increase in crystal density. For a film obtained using diethylamine this index was 22 mV, and using ice acetic acid - 8 mV, Table 1. Such a substantial increase can be caused by a decrease in deficiency of crystals as ultrasonic modifying promotes formation of dense crystal package. Recombination of photoelectrons and holes, apparently, is the main reason of a decrease in photocatalytic activity of the materials obtained without ultrasonic treatment. Emergence of p-type conductivity, apparently, is determined by non-stoichiometry and occurrence of discrete levels in the bandgap because of an excess of acceptor impurities in the built crystal lattices formed as a result of formation of hybrid compounds. Table 2 summarizes the results of comparative characteristics of the photoresponse films.

Material designation	Polarity of conductivity	Electrode	Photo EMF, mV	Average crystallite size, nm
TiO_2, pH=4	P-	Ni	1.5	7.5
TiO_2, pH=11	P-	Ni	6.8	6
TiO_2+ USI, pH=4	P-	Ni	8.0	9
TiO_2+ USI, pH=11	P-	Ni	22.0	8.1

Table 1. The response of photoelectromotive force.

9. Anodic oxidation

As compared to the other methods of obtaining the corresponding oxides on the metal surface (thermal, chemical oxidizing), electrochemical oxidation has a number of advantages. In particular, anodic oxidation is one of the most convenient and simple ways of obtaining thin oxide films in non-equilibrium conditions with formation of metastable structural and chemical phases displaying required properties [Alphonse et al., 2010; Wang et al., 2009].

For the realization of the process of electrochemical titanium oxidizing, using a simple electric chain is enough (Fig. 5). The electrochemical cell represents the two-electrode system consisting of the titanium anode and the corrosion-resistant steel cathode connected to the direct current source.

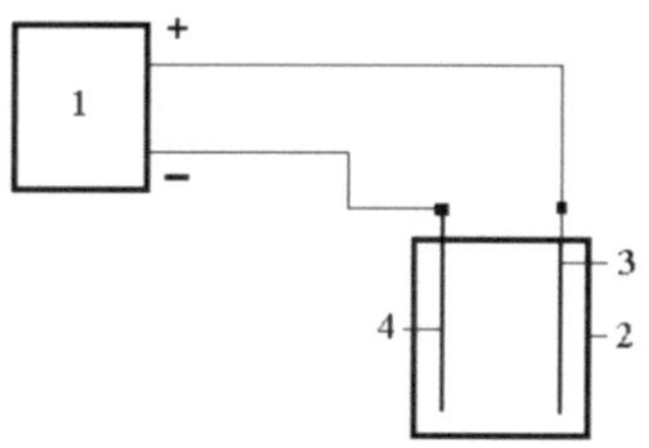

Fig. 5. The block diagram of a cell for titanium anodization process: 1 – power supply; 2 – electrochemical cell; 3 – anode; 4 – cathode.

Titania-based films obtained using electrochemical oxidation of metallic titanium possess an interesting microstructure. During anodic oxidation on the titanium surface the porous titania film is formed, consisting of a mass of nanotubes oriented perpendicularly to metal substrate whose diameter can be varied within several tens of nm depending on parameters of electrochemical process [Gong et al., 2001; Macak et al., 2005; Paulose et al., 2007]. The choice of suitable modes of anodic oxidation allows to obtain porous titania films with the required pore sizes and good homogeneity.

To determine the effect of solvent nature on the kinetics of anodic process, we have recorded the chronoamperometric curves for EG and DMSO-based electrolytes (Fig. 6).

In chronoamperograms the initial part of titanium anodizing process is reflected. Initial current density in EG-based electrolytes is twice as much compared to DMSO-based ones. It can be attributed to the different electrical conductivities of the given solutions. The specific electrical conductivity of EG-based electrolyte is $8{,}72 \cdot 10^{-3}$ S/m whereas that for DMSO-based solution is $3{,}97 \cdot 10^{-3}$ S/m.

In the DMSO-based solutions upon exposure to direct voltage the current decline occurs at a smaller speed as compared to EG-based electrolytes that testifies to the formation of an oxide layer with more advanced surface. In EG-based electrolytes a sharp current decline is observed, that testifies to the formation of titania layer in a shorter time.
Photoelectrochemical properties of the surface layers generated on titanium by anodic oxidation were investigated in background electrolyte 0.2M Na_2SO_4. The photoresponse with negative sign testifies to the formation on the titanium surface of a non-stoichiometric oxide with lack of oxygen and electronic type of conductivity. For the purpose of increasing the photoresponse of titania films obtained using anode oxidation, we additionally performed thermal treatment directly after anodizing process.

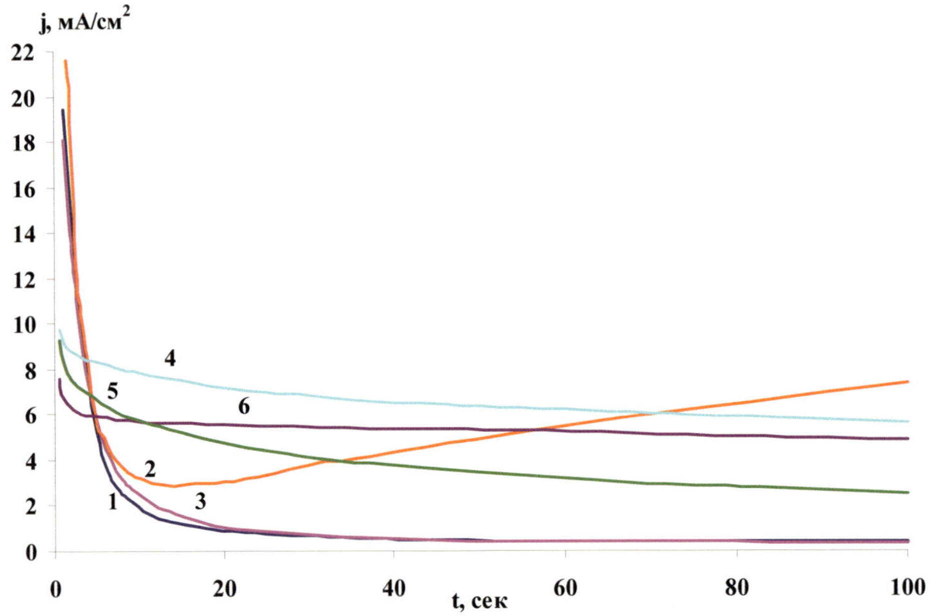

Fig. 6. Chronoamperometric curves of titanium electrode in electrolytes: 1 - EG + 2.5 g/l NH_4F; 2 - EG + 2.5 g/l NH_4F + 0.15 g/l PcCo; 3 - EG + 2.5 g/l NH_4F + 0.15 g/l $CoSO_4 \cdot 7H_2O$; 4 - DMSO + 0.5 g/l NH_4F + 5% H_2O; 5 - DMSO + 0.5 g/l NH_4F + 5% H_2O + 0.15 g/l PcCo; 6 - DMSO + 0.5 g/l NH_4F + 5% H_2O + 0.15 g/l $CoSO_4 \cdot 7H_2O$.T = 293 K. Cell voltage is 20 V.

For increasing the photocatalytic properties of titania films we have chosen the higher macroheterocyclic compounds (porphyrins and phthalocyanines), and also their metal-containing complexes. Metal-containing porphyrin and phthalocyanines complexes possess unique optical, semiconductor and catalytic properties, therefore are of interest for researchers in the field of chemistry, biology, medicine, optics and materials technology.
It is seen from the Table 1 that the greatest photo-EMF magnitudes are exhibited by titania films obtained in DMSO solution containing 0.15 g/l of cobalt phthalocyanine, and that is apparently related to introduction of the latter into porous structure of the oxide layer.

Solution composition	V_{PEP}, mV for thermal treatment time, min		
	0	30	60
EG+2.5 g/l NH_4F	-1	-30	-32
EG +2.5 g/l NH_4F + 0.15 g/l $CoSO_4 \cdot 7H_2O$	-8	-60	-61
EG +2.5 g/l NH_4F + 0.15 g/l PcCo	-7	-58	-59
DMSO + 5 % H_2O + 0.5 g/l NH_4F	-10	-61	-62
DMSO + 5 % H_2O + 0.5 g/l NH_4F + 0,15 g/l $CoSO_4 \cdot 7H_2O$	-10	-61	-62
DMSO + 5 % H_2O + 0.5 g/l NH_4F + 0,15 g/l PcCo	-20	-68	-68

Table 2. The photoelectropolarization (V_{PEP}) magnitudes for a titanium electrode after anodization within 60 minutes in DMSO solutions with addition of Co-containing salts and thermal treatment at T = 523 K.

The photo-EMF magnitudes for titania films obtained within 1 hour of anodic oxidation in solutions with different concentrations of PcCo, are listed in Table 3. It is seen from the table that the photoelectropolarization magnitudes directly after anodization process range from 20 to 40 mV, the introduction of PcCo into solution leading to an increase in the photo-EMF magnitudes, and that is related to an introduction of PcCo into the structure of the oxide layer. After thermal treatment the photo-EMF magnitudes increase. The increase in the photo-EMF magnitudes can also be related to the increase in activity of PcCo as a result of thermal treatment.

Solution composition	V_{PEP}, mV for thermal treatment time, min		
	0	30	60
DMSO + 0.05 g/l NH_4F + 5% H_2O	-24	-40	-60
DMSO + 0.05 g/l NH_4F + 0.05 g/l PcCo + 5% H_2O	-40	-63	-64
DMSO + 0.05 g/l NH_4F + 0.15 g/l PcCo + 5% H_2O	-41	-62	-68

Table 3. The photoelectropolarization (V_{PEP}) magnitudes for a titanium electrode after anodization within 60 minutes in solutions with different concentrations of PcCo and thermal treatment at T = 523 K.

We have also investigated the effect of additives of metal-containing porphyrin and phthalocyanines complexes of various nature. In Table 4 we list the photo-EMF magnitudes for titania films obtained from solutions with addition of macroheterocyclic compounds of various composition – cobalt phthalocyanine (PcCo), cobalt porphyrin (PhCo) and deuteroporphyrin (DtPh). It is notable that the greatest photoactivity is exhibited by films obtained from solutions with addition of PcCo.

Solution composition	V_{PEP}, mV for thermal treatment time, min		
	0	30	60
DMSO + 0.5 g/l NH_4F + 5% H_2O + 0.05 g/l PcCo	-40	-63	-64
DMSO + 0.5 g/l NH_4F + 5% H_2O + 0.05 g/l PhCo	-1	-61	-61
DMSO + 0.5 g/l NH_4F + 5% H_2O + 0.05 g/l DtPh	-2	-57	-60

Table 4. The photoelectropolarization (V_{PEP}) magnitudes for a titanium electrode after anodization within 60 minutes in solutions with different additives before and after thermal treatment at T = 523 K.

Introduction of the cobalt phthalocyanine additive into DMSO-based anodizing electrolyte leads to a decrease in the rate of anodic titanium oxidation. Titania layers formed during the process exhibit increased photo-EMF magnitudes. Oxide structures obtained using the method of anodic titanium oxidation in DMSO-based solutions with the cobalt phthalocyanine additive can be usefully applied in photoelectrochemical processes.

10. Sensitization of metal-oxide electrode surfaces using dyes

As the own absorption region of titanium dioxide ($\lambda < 380$ nm, $h\nu \geq 3.2$ eV) is affected by less than three percent of solar irradiation, it is sensitized to visible region. One of the ways to increase the efficiency of excitation process for wide-zone semiconductors is chemical or physical sorption of dyes. Expansion of region of operating light is possible at the expense of dye excitation by long-wave light and then by the charge transfer to the semiconductor, Fig. 7.

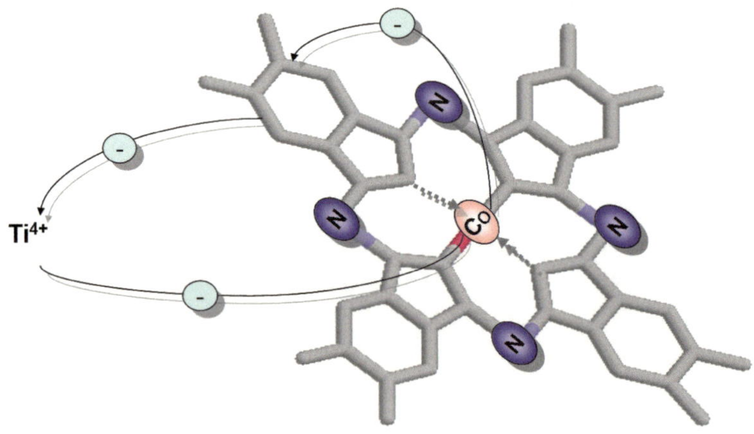

Fig. 7. An example of formation of a thin-film PcCo complex to an efficient electron acceptor - semiconductor oxide particle.

If distributions of the excited dye levels are intersected with zones of resolved energies of the semiconductor - electrode, and there is no non-excited one, then electronic transition between the electrode and the solution, i.e. the course of an electrochemical reaction, is possible only upon irradiation.

In our investigations [Vinogradov et al., 2008] modifying sol-gel synthesis by introduction of water-soluble PcCo as an organic sensitizer in the stage of formation of titania sol particles have also allowed to increase photoactivity of formed hybrid nanomaterials. In this work, using the given photopolarization measurements it has been shown that photocatalytic activity for an organic-inorganic composite titania-PcCo obtained by sol-gel technology depends on conditions of material synthesis and can be considerably higher than for a pure titania, see Table 5.

Substance	*T*, °C thermal fixation	Electrode	Photo-emf of electrode without deposited composite	Photo-emf, mV	Gain in photo-emf
TiO_2, pH 4	250	Ni	0.44	1.5	3.4
TiO_2, pH 11	250	Ni	0.34	6.8	20
PcCo + TiO_2, pH 4	250	Ni	0.44	7	15.9
PcCo + TiO_2, pH 5	250	Cu	6	9	1.5
PcCo + i-I	180	Ni	0.44	0.7	1.6
PcCo + TiO_2, pH 11	180	Ni	0.34	0.5	1.47
TiO_2, pH 11	180	Ni	0.34	0.74	2.18

Table 5. Photo-emf response of the films under UV irradiation.

11. Doping of organized semiconductor by introducing metal nanoparticle layer

Putting a metal layer on a semiconductor for the purpose of increasing its photoactivity is well-known and successfully used technique. Metal is put either in the form of a continuous thin film to let the radiation pass through the metal - semiconductor interface, or in the form of nanoclusters or large-pore films on the semiconductor surface, assuming its direct contact to electrolyte.

Precipitation of precious metals onto the semiconductor surface, as a rule, substantially changes the properties of the surface and thereby strongly affects the photocatalytic process with its participation. Metal can change composition of products or rate of photocatalytic reaction. An increase in the reaction rate was for the first time observed in the case of Pt/TiO_2 during reactions of water decomposition into oxygen and hydrogen. In [Gnaser et al., 2004] it is said that Pt, Pd, Rh increase the rate of hydrogen production by 20–100 times whereas putting iron suppresses hydrogen production.

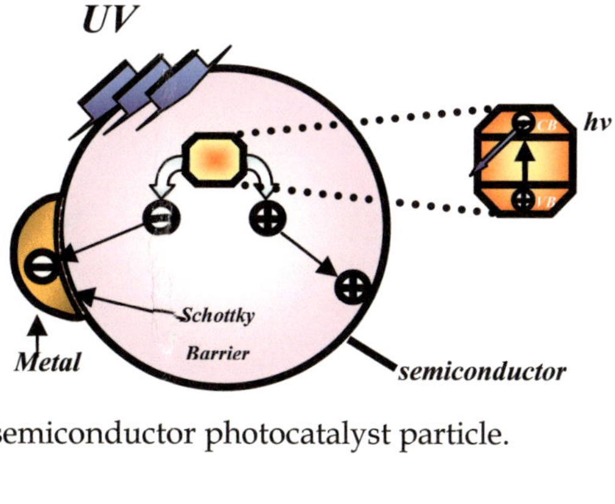

Fig. 8. Metal-modified semiconductor photocatalyst particle.

After excitation, an electron migrates towards metal by which it is grabbed, so electron-hole recombination is suppressed. Migration of electrons towards metal particles is confirmed by investigations showing a decrease in the semiconductor photoconductivity upon putting Pt on TiO_2 (as compared to pure TiO_2). But holes appear to be free enough to diffuse towards semiconductor surface, and then to enter reactions, for example, oxidations of organics. In practice the Pt/TiO_2 system is especially widely used. Platinum introduction onto the TiO_2 surface appears to be especially efficient for photocatalytic reactions in which gas is produced, in particular, hydrogen.

Along with doping by precious metals a special attention is recently paid to modifying the photocatalysts by the f-element impurities [Mazurkiewicz et al., 2005]. As the investigations reveal, all f-elements can be divided into two groups: Nd, Pm, Gd, Ho, Er, Lu which exhibit valence (III), and Ce, Pr, Sm, Eu, Tb, Dy, Tm, Yb, demonstrating variable valence (II), (III), (IV). According to the obtained data, the most efficient impurities are Pr, Sm, Eu, Dy, and Tm, i.e. f-elements with variable valence.

Apparently, upon absorption of UV-light quanta there is an increase in concentration of paramagnetic Ti^{3+} ions at the expense of free electrons: $Ln^{3+} + h\nu = Ln^{4+} + \bar{e}$. Besides, among the entire row of f-elements it is necessary to note Tm^{3+} to be the most efficient TiO_2 activator. According to the principle of lanthanide contraction, Tm^{3+} has the least ionic radius. Thus, the Tm^{3+} ions regarding the spatial-energetic relation possess higher probability to penetrate into the TiO_2 layer and to act as electron donors or the impurity adsorption centers, i.e. both collective and individual factors of affecting the state of the photocatalyst surface take place.

Modifying titania by certain impurities and preliminary thermal treatment in a reductive environment allows, within certain frameworks, to control its photocatalytic activity.

One of the goals of our investigations started in 2010 was modifying the surface of the TiO_2 film by various metals in colloidal state, such as Cu, Ag, Fig.9, which would allow, on the one hand, to achieve photochromic effect without the traditional usage of platinum and palladium, and, on the other, to increase the photoactivity of the generated composites.

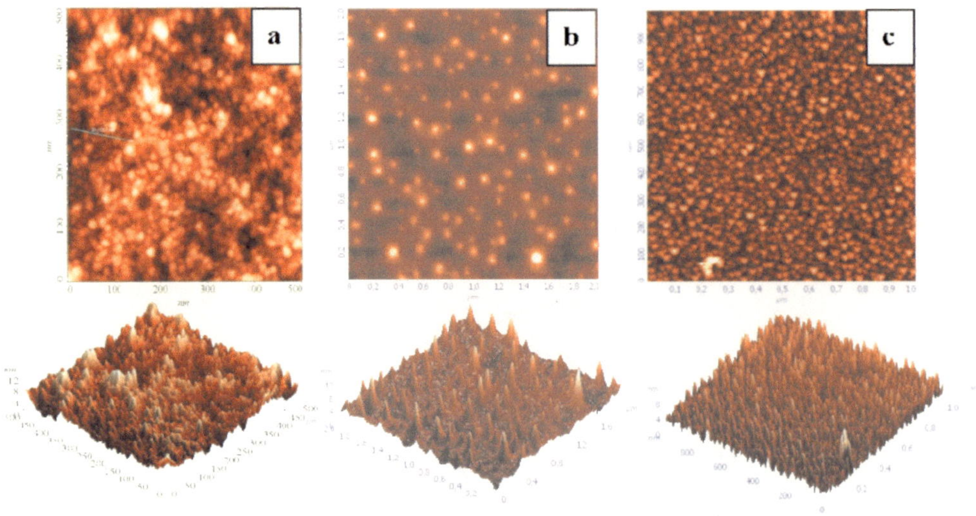

Fig. 9. The AFM images: a) the surface of the TiO_2 film; b) the surface of the Cu–TiO_2 film; c) the surface of the Ag–TiO_2 film.

Thus, the necessary factor was obtaining the nanoparticles with sizes in the range of 10 nm, leading to the generation of heterostructures of the metal nanoparticles plasmon resonance in the system. As a result of processing the AFM images, it was revealed that the copper nanoparticles generated in the conditions of the Cu^{2+} reduction using sodium tetrahydroborate, Fig. 9b, possess spherical form and narrow particle size distribution of about 8 nm. According to Fig.9(c), the generated Ag nanoparticles represent triangular nanoprisms with the average size of about 5 nm distributed in regular intervals on the TiO_2 surface.

Sample	Photo-emf (non-calcined), mV	Conductivity type
TiO_2+Ag	46	p-
TiO_2+Cu	32	p-
TiO_2 non-modified	15	n-

Table 6. The photoactivity characteristics of composite materials obtained using metal nanoparticles.

The data implying high photoactivity of the Ag/TiO_2 composite are confirmed by the greatest photoresponse value of 46 mV, Table 6. After UV irradiation the excited electrons move towards the TiO_2 conductivity zone, and holes move towards the valence zone through interface. Thus, the separation of the photogenerated electron–hole pairs in a composite film is more efficient than in pure, non-modified one. Hence, the recombination of the photogenerated charge carriers also proceeds more efficiently, and that is proved by an increase in the photoresponse value for the composites used. The use of copper and silver nanoparticles also promotes the increase in photocatalytic activity of a film, Table 6, owing to larger water adsorption on the surface of a composite due to the nanoparticle surface effect [Agafonov et al., 2009], which is promoted by a high concentration of the photogenerated holes whose presence is confirmed by composite conductivity type, Tab.6.

12. Using template synthesis for obtaining materials with high photoactivity

The analysis of literature data has shown that for studying photoactivity of titania-based materials the spectroscopic investigations aimed at studying photochemical reactions on the surfaces of materials are the most used. However, such an approach allows to only partially describe the processes of charge transport and effects of internal structure on photoactivity of a material as a whole. As has been shown, the greatest complications arise in the presence of considerable structural impurities in the lattice structure, and also in the presence of a bulky structured surface. The presence of highly electronegative elements in crystal structure, which act as electron donors, considerably complicates the process of charge transport, and such materials will be characterized, as a rule, by either hole or ionic type of conductivity. Thus it is necessary to perform a complex estimation of photoactive properties of synthesized materials which would allow to consider simultaneously the role of structure and nature of a material. Using template synthesis, on the one hand, allows to form various highly arranged structure of materials in mesoregion and, on the other, in the course of removing templates by thermal treatment, leads to an increase or a decrease in photoactivity (depending on type of structure of generated crystallites) because of remaining presence of impurity ions, such as C, N, O etc. Thus, the most fair is the use of a combination of two methods - method of photoelectric polarization of films and analysis of kinetic curves of a model dye photodestruction which provide the accounting for both the effect of the structural factor and own semiconductor properties on the total photocatalytic properties.

The most interesting study is supposed to be that of functional properties of the preparations obtained using templates which differ by their chemical nature: dodecylamine (DDA), polyethylenimine (PEI), polyethylene glycol monooleate (PEGMO), polyethyloxazoline (PEOA). The structure of hybrids obtained with their participation is shown in Fig. 10.

It is seen from the presented figures that using various modifying additives leads to various organizations of the surface. In Fig. 10a the structure of the TiO_2 film formed by hierarchical pores of the roundish shape (Ø ≈ 105 nm) with uniform morphology is shown. The films obtained using dodecylamine, are characterized by pores with the narrowest size distribution (Ø ≈ 30 nm). The films generated with participation of PEGMO, are covered with oval pores with the maximum length of 150 nm, and the length to width ratio of about 5. It is obvious that the pore size is related to the degree of hydrophobicity of a template. The materials including hydrophilic PEI and PEGMO reveal larger pores than with hydrophobic DDA. At the same time, it has been established that for the films generated using tertiary amines, fig. 10d, the formation of planar structure is observed, with "islet" inclusions which are distinguished by a chaotic spatial organization with non-uniform formations. It points to the fact that the coordination activity of the stabilizer is low. Thus, for the films obtained as a result of isopropylate hydrolysis in the presence of polyethyloxazoline we can observe separate formation of large agglomerates of hydrated titania and planar structures of polymer - polyethyloxazoline (fig. 10d).

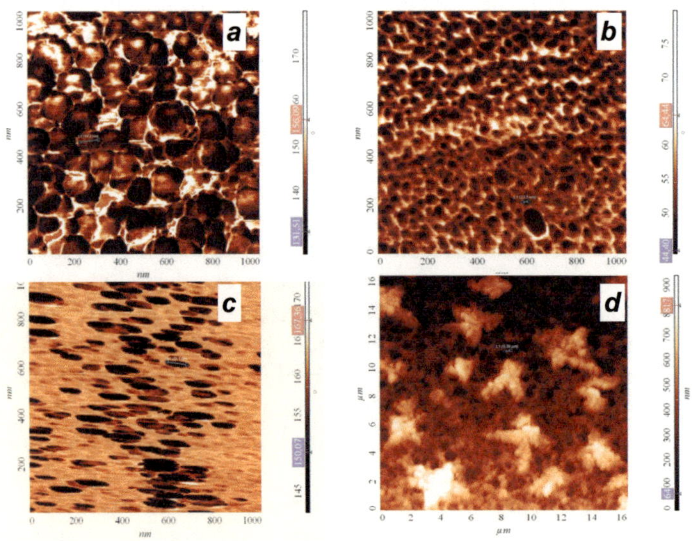

Fig. 10. The surfaces of hybrid films modified by a) polyethylenimine; b) dodecylamine; c) polyethylene glycol monooleate; d) polyethyloxazoline.

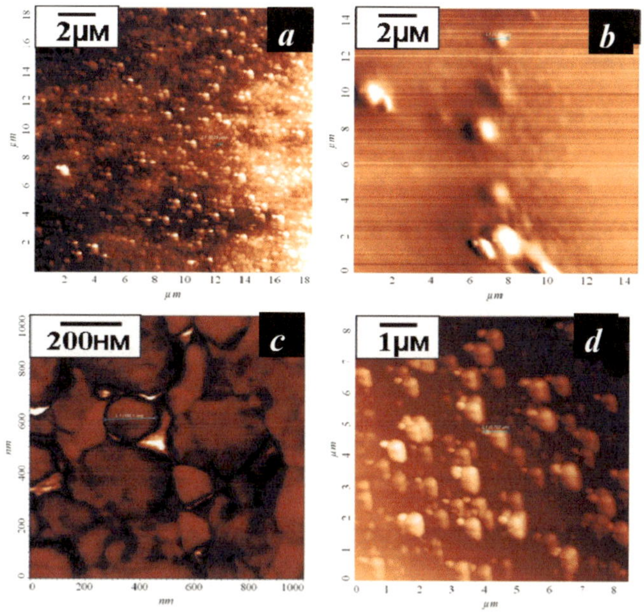

Fig. 11. The surfaces of calcined (at 300°C) films modified by: a) polyethylenimine; b) dodecylamine; c) polyethylene glycol monooleate; d) polyethyloxazoline.

Among the obtained materials, the greatest gain in photo-EMF is exhibited by the calcined films generated in the presence of polyethyloxazoline and polyethylene glycol monooleate - templates with the least coordination activity leading to the formation of defective crystallites and n-type conductivity.

The primary and secondary amines which are characterized by high coordination activity, during synthesis promote formation of stable inorganic frameworks in which nitrogen is chemically bonded to Ti^{4+}. After calcination in such materials the accepting impurity of nitrogen remains, which forms additional energy levels. It is necessary to note that using polyethylenimine as a template allowed to generate films with high photoactivity, both in non- calcined (2.8 mV), and in calcined form (20 mV), due to the presence of a larger number of developed electron accepting impurity centers that promote narrowing of the bandgap, and, as a consequence, a "readier" formation of electron-hole pair. Comparison of photoactivity of thin TiO_2-based coatings obtained using various methods allows to make a conclusion about the prospects of using modifying additives for the purpose of increasing the quantum yield, as in this case photogenerated electron-hole pairs in the TiO_2 nanoparticles possessing short-range order are separated more efficiently than in pure TiO_2. Thus, the excess of accepting impurity in the structure of crystal lattice can delay the recombination of photogenerated electrons and holes and thereby promotes increase in the TiO_2 photocatalytic activity.

The film photoactivity evaluation was performed using photo-emf data upon the brief irradiation by a 250 W UV lamp; a platinum screen served as the second electrode. The data obtained are listed in Table 7.

TiO_2 is known to be indirect bandgap semiconductor characterized by electronic conductivity type. This charge transport mechanism is due to the formation of O^{2-} vacancies in the crystal lattice structure, the two neighbor Ti^{4+} ions acquiring the $^{3+}$ charge. It leads to the appearance of a weakly connected electron on their outer electron shell bringing about the conductivity type. The presence of highly electronegative elements acting as electron donors in crystalline structure significantly impedes the charge transport process, and such materials will as a rule be characterized by hole or ionic conductivity type.

Modified TiO_2 sample	Surface pore diameter, nm	Average crystallite size	V, cm^3/g	D_{pore}, Å	Conductivity type	Photo-emf, mV
Polyethyloxazoline	112	2	0.035	117	N-	45
Polyethylenimine	105.2	2.5	0.584	282	P-	20
Dodecylamine	49.3	1.7	0.174	58	P-	1.5
PEG monooleate	17.8-92.4(L)	2.1	0.265	110	N-	22.5

Table 7. The resulting table of physico-chemical and structural properties.

The assumption of low coordination activity of polyethyloxazoline is also confirmed by the data from Table 7. For such a film, the greatest photo-EMF of 45 mV caused by the formation of the least defective crystals and n-type conductivity is found. For comparison, the characteristic of TiO_2 film modified by PEG monooleate is given, indicating the low coordination activity in complexation reactions. The data obtained show that the both films possess high photoactivity and n-type conductivity.

The primary and secondary amines characterized by high coordination activity promote the formation of stable inorganic frameworks during the synthesis process. In these frameworks nitrogen is chemically bonded to Ti^{4+}. After calcination these materials retain the acceptor impurity of nitrogen that forms additional energy levels. At the same time, using the polymer molecule, i.e. polyethylenimine as a template promotes the formation of films with highly developed surface and photoactivity due to the presence of a large number of developed electron acceptor groups. In the diethylamine-octylamine-dodecylamine series the photoactivity decreases drastically as the basicity of an amine decreases.

13. Conclusion

Thus, using method of photoelectric polarization for estimating the functional properties of the materials used upon manufacturing photocatalysts and solar cells acting on the basis of the modified solid-state semiconductors is the most universal and readily available technique. In the given chapter we have shown the basic ways promoting the increase in photoactivity of titania-based materials obtained by anodic oxidation and sol-gel method. Among those are using ultrasonic treatment, modifying by phthalocyanines and metal nanoparticles, formation of highly developed surface, doping with metals and non-metals. The fundamental aspects of the photo-EMF emergence in nanomaterials upon electrode irradiation in a solution in combination with the resulted data which have been considered in this chapter allow to use competently the described method and apply it for estimating the photoactivity of materials.

14. Acknowledgments

This work was supported by the Russian Foundation for Basic Research, Projects No. 09-03-97553, 11-03-12063, 11-03-00639, 10-03-92658.

15. References

[1] Agafonov, A.V., Vinogradov, A.V.(2009). Sol–gel synthesis, preparation and characterization of photoactive TiO_2 with ultrasound treatment. *J Sol-Gel Sci Technol.*, 49, pp. 180–185.

[2] Agafonov, A.V., Vinogradov, A.V.(2008). Catalytically Active Materials Based on Titanium Dioxide: Ways of Enhancement of Photocatalytic Activity. *High Energy Chemistry*, 42, pp.70–72.

[3] Alphonse, P., Varghese, A., Tendero C.(2010). Stable hydrosols for TiO2 coating, *J Sol-Gel Sci Technol*, 56,pp. 250–263.

[4] Gnaser, H., Huber, B., Ziegler C.(2004). Nanocrystalline TiO_2 for Photocatalysis. *Encyclopedia of Nanoscience and Nanotechnology*, 6, pp.505–535.

[5] Gong, D., Grimes, C. A., Varghese, O. K., Hu, W., Singh, R.S., Chen, Z., Dickey, E. C.(2001) Titanium oxide nanotube arrays prepared by anodic oxidation. *J. Mater. Res.*, 16(12), pp. 3331-3334.

[6] Grätzel, M., O'Regan, B., (1991). A low-cost, high-efficiency solar cell based on dye-sensitized colloidal TiO_2 films. *Nature* , 353 (6346), pp. 737–740.

[7] Kityk, I.V., Sahraoui, B., Fuks, I., et al.(2001) Novel nonlinear optical organic materials: dithienylethylenes, *J. Chem. Phys*, 115(13),pp. 6179-6184.

[8] Kwong, C. Y., Choy, W. C., Djurisc, A. B., Chui, P. C., Cheng, K. W. and Chan, W. K.(2004). Poly(3-hexylthiophene):TiO2 nanocomposites for solar cell applications, *Nanotech.* 15,pp. 1156-1161.

[9] Li, Y., Hagen, J., Schaffrath, W., Otschik, P., and Haarer, D.(1999). Titanium dioxide films for photovoltaic cells derived from a sol-gel process, *Sol. En. Mat. Sol. Cells*, 56,pp. 167-174.

[10] Macak, J. M., Tsuchiya, H., Schmuki, P.(2005). High-Aspect-Ratio TiO_2 Nanotubes by Anodization of Titanium. *Angewandte Chemie International Edition*, 44 (14), pp. 2100 - 2102.

[11] Masakazu, A.(2000), Utilization of TiO_2 photocatalysts in green chemistry, *Pure Appl. Chem.* 72, pp. 1265 -1270.

[12] Mazurkiewicz, J.S., Wlodarczyk, R.P., Mazurkiewicz, G.J.(2005). Effect of f-elements on photocatalytic activity, electrical conductivity and magnetic susceptibility of titanium dioxide, *Chemistry and chemical technology*, 48(1), pp. 118-121.

[13] Osche, E.K., Rosenfeld, I.L.(1969). Method of photoelectric polarization for studying the deviation from stoichiometry of surface oxides on metal electrodes, *Protection of metals.*, 5(5),pp. 524-531.

[14] Osche, E.K., Rosenfeld, I.L.(1978). Scientific and technical results: Corrosion and corrosion protection. *Protection of metals*, 7,pp. 111-158.

[15] Pai, R. R,. John, T., Kashiwaba,Y., Abe, T., Vijayakyumar, K.P., Kartha, C. S., (2007) Photoelectrical properties of crystalline titanium dioxide thin films after thermo-annealing, *J. Mat. Sci.* 42(5), pp.498-503.

[16] Paulose, M., Prakasam, H. E., Varghese, O. K., Peng, L., Popat, K.C., Mor, G. K., Desai, T.A. and Grimes, C. A. (2007) TiO_2 nanotube arrays of 1000 μm length by anodization of titanium foil: phenol red diffusion, *J. Phys. Chem. C*, 111(41), pp. 14992–14997.

[17] Rincon, M. E., Daza, O., Corripio, C., Orihuela, A.(2001) Sensitization of screen-printed and spray-painted TiO2 coatings by chemically deposited CdSe thin films. *Thin Solid Films*, 389, pp.91-98

[18] Vakalov, D. S., Rydanov, R. S., Bairamukov, O. M., Krandievsky, S. O. Ilyasov, A. S., Mikhnev, L. V.(2010). Study on optical and photoelectrical properties of powder zinc. *Bulletin of North Caucasus State Technical University*.3 (24),pp. 46-49.

[19] Vinogradov, A.V, Agafonov, A.V, Vinogradov, V.V.(2009). Sol-gel synthesis of titanium dioxide based films possessing highly ordered channel structure, *J. Mendeleev Comm.*, 19, pp.340-341.

[20] Vinogradov, V.V, Agafonov, A.V., Vinogradov, A.V.(2010). Superhydrophobic effect of hybrid organo-inorganic materials. *J Sol-Gel Sci Technol.*, 53,pp. 312–315.

[21] Wang, J., Lin, Z.(2009) Anodic formation of ordered TiO_2 nanotube arrays: Effects of Electrolyte Temperature and Anodization Potential, *J. Phys. Chem. C.*, 113(10), pp. 4026-4030.

[22] Weidmann, J., Dittrich,T., Konstantinova, E., Lauermann, I., Uhlendorf, I., Koch, F., (1998) Influence of oxygen and water related surface defects on the dye sensitized TiO_2 solar cell, *Sol. En. Mat. Sol. Cells* 56, 153-165.

[23] Zeenath, N. A., Pillai, P. K. V., Bindu, K., Lakshmy, M., Vijaya Kumar, K. P. (2000) Study of trap levels by electrical techniques in p-type $CuInSe_2$ thin films prepared using chemical bath deposition, *J. Mat. Sci.* 35,pp. 2619-2624.

3

Electromotive Force in Electrochemical Modification of Mudstone

Dong Wang[1,2], Jiancheng Song[1] and Tianhe Kang[1]
[1]*Taiyuan University of Technology, Taiyuan,*
[2]*Shanxi Coal Transportation and Sales Group Co.Ltd, Taiyuan,*
China

1. Introduction

It is utilized in the coal-mine soft rock roadway that bolt with wire mesh, grouting and guniting combined supporting technique and quadratic supporting technique. The supporting techniques can anchor high stressed soft rock and jointed soft rock, however, with little help for mudstone. The analyses of deformable mechanism in mudstone roadway are based on engineering mechanical property of mudstone, which mainly includes swelling and disintegration. On the other hand, the mineralogical composition of mudstone is quartz, calcite, montmorillonite, illite, kaolinite, and chlorite. The analyses lead to the following conclusion: engineering mechanical property of mudstone induced by the shrink-swell property of clay minerals, swelling clay minerals play significant roles in the swelling process of mudstone.

In swelling clay minerals there are two types of swelling. One is the innercrystalline swelling caused by the hydration of the exchangeable cations of the dry clay; the other is the osmotic swelling resulted from the large difference in the ion concentrations close to the clay surfaces and in the pore water. The swelling of clay minerals as it manifests itself in the coal-mine mudstone roadway is referred to as the osmotic swelling.

The electrochemical modification of clay minerals is that the electrodes and the electrolyte solutions modify clay minerals under electromotive force, leading to change in the physical, chemical and mechanical properties of clay minerals. Electrochemical modification of clay minerals was applied in soil electrochemistry (Adamson et al., 1967; Harton et al., 1967; Chukhrov, 1968; Gray, 1969), electrical survey (Aggour & Muhammadain, 1992; Aggour et al., 1994), stabilization of sedimentary rock (Titkov, 1961; Titkov., 1965), and mineral processing (Revil & Jougnot, 2008). According to the applications, the mechanism of electrochemical modification of clay minerals is summarized as follow (Adamson et al., 1966; Harton et al., 1967):

- electroosmotic dewatering and stabilization;
- cation substitutions, structures and properties change, forming new minerals.

After electromotive force treatment, the main analyses of properties centralize into the physicochemical and mechanical properties. Physicochemical and mechanical properties of mudstone changed through electrochemical modification, the modified purpose to change other unfavorable properties of mudstone, such as mechanical property (uniaxial

compressive strength, tensile strength, and triaxial compressive strength) and engineering mechanical property (plasticity, swelling, rheology and disturbance characteristics).
With respect to the modification of mudstone by electrochemical method, the essence of the method is electrochemical modification of physicochemical properties of clay minerals. It is our destination task that the conventional electrochemical stabilization of clay minerals may be applied to support mudstone roadway in coal-mine.

2. Electrochemical dewatering and stabilization

Under electromotive force treatment, electrochemical dewatering and stabilization is based on the electrically induced flow (namely, electroosmosis) of water trapped between the particles of clay minerals. Such electrically induced flow is possible because of the presence of the electrical double layer at the solid/liquid interface.

2.1 Electroosmosis and electrolysis phenomenon

Electroosmosis is the motion of ionized liquid relative to the stationary charged surface by an applied DC fields. It should be emphasized that electroosmotic dewatering is most attractive when the water is trapped between fine-grained clay particles.
In 1808 the discovery of electroosmosis phenomenon (Amirat & Shelukhin, 2008) by Reiss occurred soon after the first investigations on the electrolysis phenomenon of water by Nicholson and Carlisle. Reiss observed that a difference in the electric potentials applied to the water in a U-tube results in a change of water levels (Fig.1) when the tube is filled partially with thin sand.

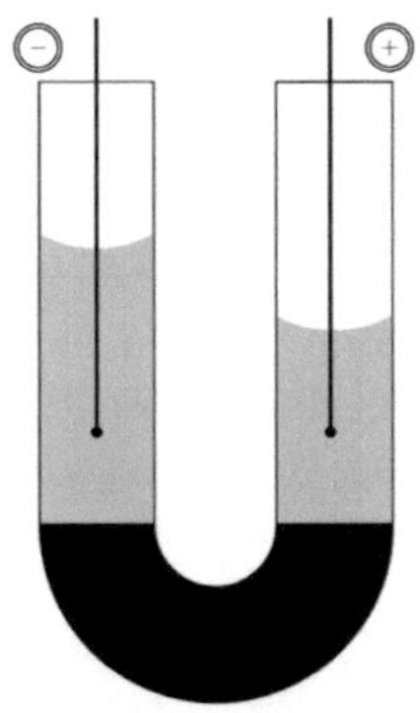

Fig. 1. Electroosmosis (Amirat & Shelukhin, 2008).

According to the surface charge properties of the clay minerals, fine-grained clay particles present in sedimentary rock normally net negative electric charges, whereas groundwater is the electrolyte solutions in nature. On the surfaces of fine-grained clay particles there exists an excess of negative charges, forming the electrical double layer. The inner or Stern layer consists of negative ions adsorbed onto the solid surface through electrostatic and Van Der Waals' forces, the ions and the oppositely charged ions in the absorbed layer do not move. The outer diffuse or Gouy layer is formed by oppositely charged ions under the influence of ordering electrical and disordering thermal forces, the positively charged ions can move.
In the presence of electromotive force in conjunction with addition of the electrolyte solutions, the electrical conductivity of clay soils increases. The assumption is as follows:

an external electric field is parallel to the solid-liquid interface in the capillary. Positive ions being formed in great quantities by the action of the electric current move in the direction of the cathode and carry with water molecules to which they are attached. The velocity of the electrolyte solutions in the electrical double layer is described by the relationship:

$$v=\varepsilon E\zeta/\eta \qquad (1)$$

where ε is the dielectric constant; E is the electromotive force; ζ is the zeta potential as the potential difference in the electrical double layer; η is the viscosity of the electrolyte solutions.

The electroosmotic velocity under the unit electric field intensity can be written as:

$$v_e=v/E=\varepsilon\zeta/\eta \qquad (2)$$

In the capillary, the thickness of the electrical double layer is negligible with respect to the capillary radius, most of the fluid in the capillary moves with a velocity. The electroosmotic velocity can be given by:

$$v_e=K_e\partial E/\partial L \qquad (3)$$

where $K_e= \varepsilon\zeta/4\pi\eta$ is the electroosmotic coefficient; $\partial E/\partial L$ is the electromotive force gradient; L is the distance between the two electrodes.

Fine-grained clay particles are negatively charged mostly because of cation substitutions. The charge is balanced by exchangeable cations adsorbed to the surfaces of clay minerals. The internal balance of charges is incorporated in the electrical double layer. Potassium and sodium cations contained in the outer diffuse layer are substituted by electrically stronger hydrogen, calcium, and aluminum cations. The substitution leads to a decrease in the thickness of water film on the clay particles and to a considerable decrease in hydrophilic tendency of the clays. Thus, the size of some of the clay particles decreases. Decrease in size and charge of the particles results in coagulation, crystallization, and adsorption of small particles on the surfaces of the larger ones. Coagulation and crystallization are very important in the whole electroosmotic processes.

During the electroosmotic processes, the electrolyte solutions in the vicinity of the electrodes are electrolyzed. Oxidation occurs at the anode, oxygen gas is evolved by hydrolysis. Reduction takes place at the cathode, hydrogen gas evolved. The electrolysis reactions are:

$$\text{At the anode} \qquad 2H_2O-2e^-\rightarrow O_2+4H^+ \qquad (4)$$

$$\text{At the cathode} \qquad 2H_2O+2e^-\rightarrow H_2+2OH^- \qquad (5)$$

As the electrolysis proceeds, the zeta potential near the anode decreases because of the decrease in pH caused by reaction (4). Near the cathode, the pH remains high during electrolysis and changes little.

The process of the electrolysis is affected by the electromotive force, the electrolyte solution, and temperature. Dewatering and stabilization resulted in several physicochemical and chemical processes which take place concurrently, there is difficultly in evaluating the contribution of each to the effectiveness of dewatering and stabilization.

2.2 Electroosmotic dewatering and stabilization

Various structural clay minerals exhibit significant differences in substitute mechanism and in the ratio between permanent and induced charges. Fine-grained clay particles have negative charges resulted from ionization, ion adsorption, and cation substitutions. The main reason is cation substitutions.

The consolidation theory by Terzaghi has connected with electrochemical stabilization of clay minerals through electroosmosis. The differential equation governing the unidirectional electroosmotic consolidation can be expressed as follows (Zhang et al., 2005):

$$\partial u/\partial t = C_v \partial^2 u/\partial z^2 \tag{6}$$

where C_v is the coefficient of consolidation, $C_v = k/r_w m_v = (1+e)k/r_w a_v$; k is the coefficient of permeability; r_w is the unit weight of water; m_v is the coefficient of volume compressibility; $m_v = a_v/(1+e)$, u is the excess hydrostatic pressure ; a_v is the coefficient of compressibility.

The initial and boundary conditions for the solution of equation (6) are:

$$u|_{t=0} = u_0 \tag{7}$$

$$u|_{z=0} = -r_w V K_e / K_h \tag{8}$$

$$\partial u/\partial z|_{z=H} = 0 \tag{9}$$

The corresponding solution of equation (6) can be given as (Zhang & Wang, 2002):

$$u = (4/\pi) P_e \sum_{0}^{\infty} [1/(2n+1)] \sin[\pi z(2n+1)/2H] \exp(-T_v(2n+1)^2 \pi^2/4) - P_e \tag{10}$$

where $T_v = c_v t/H^2 = k_h(1+e_1)t/a_v r_w H^2$, H is the thickness of the clay layer; k_h is the hydraulic conductivity; K_e is the coefficient of electroosmotic permeability; V is the compression volume, $P_e = r_w V K_e/k_h$.

The total degree of electroosmotic consolidation defined in terms of settlement can be given by:

$$U = (4/\pi) P_e \sin(\pi z/2H) \exp(-T_v \pi^2/4) - P_e \tag{11}$$

In cation substitutions of clay minerals, the electrolyte solutions should include calcium chloride, aluminum sulfate, aluminum acetate or a mixture of several electrolytes, the anode should be aluminum electrode.

3. Modification of physicochemical and mechanical property

With respect to modification of mechanical property, the analyses of literatures lead to the following conclusions after electromotive force treatment (Adamson et al., 1966; Adamson et al., 1966; Adamson et al., 1967; Harton et al., 1967):

- The clay saturation decreased.
- Tensile load ratio values much higher than those for the materials in the natural state.
- The reduction in shrinkage crack may be considerably.
- The tensile strength and uniaxial compressive strength in mudstone increased.
- The possibility of dewatering and stabilization of clay soils by means of electromotive force. The degree of soil stabilization and course of the processes are dependent on clay content, types of clay present, and the concentration of the electrolyte solutions.

- The shrinkage of mudstone flour may be insignificant.

With respect to modification of physicochemical and mechanical property, Chilingar (Chilingar, 1970) and Aggour (Aggour & Muhammadain, 1992; Aggour et al., 1994) studied the effect of the electromotive force on the permeability of mudstone. The results are listed below:

- The permeability and wettability of cores affected by such factors as the property of the electrical double layer, the electrical conductivity of the system, the magnitude and direction of the electrical potential gradient, and the ratio of the electroosmotic to hydrodynamic water transports.
- For the mudstone full saturated with the electrolyte solutions, the greater the resistivity, the greater is the magnitude of electroosmotic transport for the same electromotive force; a linear relation exists between the applied electromotive force gradient and the electroosmotic velocity.
- During triaxial failure test, the electrokinetic coupling coefficient increased.

4. Newly-formed minerals in clay minerals and mudstone

The electroosmosis can indurate clay minerals and mudstone under electromotive force treatment. The electrolyte solutions diffuse through the clay minerals and mudstone by means of ionic transmission, changing its physicochemical properties and forming newly minerals. Titkov (Titkov et al., 1965) studied newly-formed minerals, which were formed by application of different electrodes in conjunction with the addition of electrolytes in the anodic, cathodic and intermediate zones. The electrolytes consisted of 0.1% Na2SiO3, saturated CaSO4, 1% AlCl3, FeCl2 and NaCl. The electrode materials were fabricated by aluminum, iron and graphite. Limonite was formed in the anodic zone, allophane and hisingerite were formed in the middle zone and allophane, lepidocrocite, hydrohematite and gibbsite were formed in the cathodic zone. Adamson (Adamson et al., 1967) performed electrochemical experiments on 100ml of mudstone powder, the electromotive force range from 20mA to 60mA, and found a newly-formed mineral: hisingerite. Harton (Harton et al., 1967) performed similar experiment. With electrochemical modification of mudstone powder in conjunction with the addition of an iron electrode and a 50% concentrated electrolyte of CaCl2 and Al2(SO4)3. Then they found that the newly-formed minerals were calcite, an unknown aluminum silicate, iron oxides and gypsum. Youell (Youell, 1960) applied electrochemical modification to the montmorillonite and discovered that the montmorillonite was being converted to a clay mineral with properties similar to chlorite. Sun hu(Sun, 2000) ran X-ray diffraction (XRD) analysis for clay minerals after electrochemical modification. He found that the crystal structure of montmorillonite in the anodic zone had little change, the major diffraction peak was weakened and the chlorite diffraction peak had completely vanished.

Scanning electron microscopy (SEM) and X-ray diffraction analyses lead to the following conclusions:

- In anodic zone of mudstone, sheet structures of clay minerals reduced, calcite vanished.
- The content of swelling clay minerals reduced.
- In intermediate and cathodic zones of mudstone, sheet structures of clay minerals increased, lots of quartzes exited.

5. Experimental studies

5.1 Experimental apparatus

The experimental apparatus used for the electrochemical treatment is shown schematically in Fig. 2. It mainly consists of a plexiglass pipe, electrode, the mudstone sample, electrolyte,

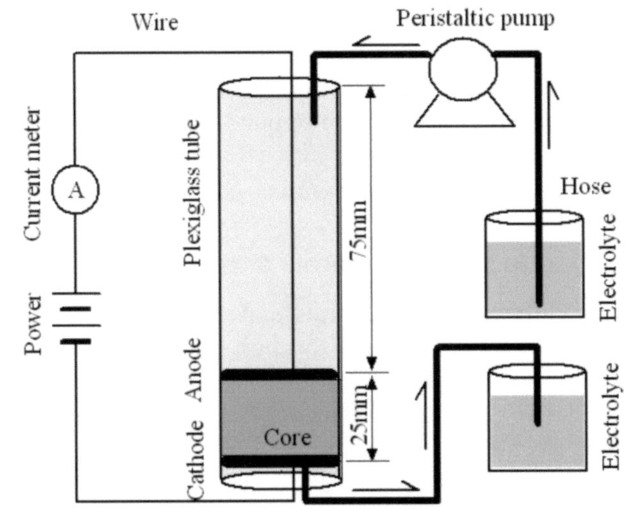

Fig. 2. Experimental apparatus.

electromotive force, current meter, peristaltic pump, hose, and wire. The electrode is a chip-type element. The anode (2 mm thick aluminium) is placed high in the plexiglass tube, whereas the cathode (0.5 mm thick red copper) is placed below the anode. The electrolyte consists of distilled water and $CaCl_2$. The electromotive force provides a voltage output ranging from 0 to 250 V and a maximum current of 1.2 A. The wire is an ASTVR $\Phi 0.35 \times 1$ mm silk-covered wire. The flow range of the peristaltic BT100-1J pump is from 0.1 rpm to 10 rpm. The pump head is an YZ1515w model. The #13 hose is $\Phi 1.6 \times 2$ mm.

5.2 Experimental sample

The specimen which taken from the roof of the 3410 tail entry of the mine at Gaoping (in the province of Shanxi, China), was a continental clastic sedimentary rock, from the Lower Permian Shanxi formation. The specimen was sealed in the tail entry, and processed into 80 cylindrical samples, each 50 mm in diameter and 25 mm in height, which were then sealed with wax in the laboratory. An example of the X-ray diffraction patterns of the samples is shown in Fig. 3. The mineralogical composition of the sample was analysed quantitatively with an adiabatic method. The mineral content of the sample was illite (45%), kaolinite (10%), quartz (38%), and anorthite (7%).

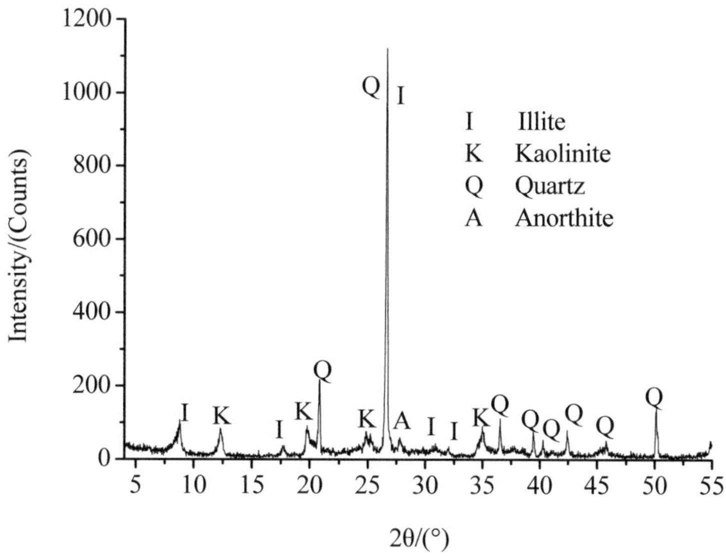

Fig. 3. X-ray diffraction pattern of experimental sample.

5.3 Experimental scheme

To investigate the tensile strength of the samples under different electrochemical treatments, 11 experimental schemes were designed, as shown in Table 1, and each scheme was applied to six samples. Scheme 1 was used to investigate the tensile strength of the original sample; scheme 2 was used to investigate the tensile strength with the power off and the sample submerged in distilled water; and schemes 3~11 were used to investigate the tensile strength under an electric gradient of 5 V cm^{-1} and at electrolyte concentrations of 0~4 mol L^{-1} (Wang et al., 2009).

Scheme	Electromotive force gradient (/V cm^{-1})	Electrolyte (/mol L^{-1})	Time modified (/h)
1	–	–	–
2	0	0	120
3	5	0	120
4	5	0.05	120
5	5	0.125	120
6	5	0.25	120
7	5	0.5	120
8	5	1	120
9	5	2	120
10	5	3	120
11	5	4	120

Table 1. Experimental schemes.

5.4 Experimental process

The samples were modified with the experimental apparatus shown in Fig. 2, according to the experimental schemes shown in Table 1. The Brazilian test, performed on a PC-style electro-hydraulic servo universal testing machine, was used to measure the tensile strength.

6. Experimental studies

Table 2 shows the measured tensile strengths of the samples. The mean tensile strength of the six original samples was 1.31 MPa in scheme 1. When the samples were submerged in distilled water, as in scheme 2, the mean tensile strength was 0.81 MPa, a reduction of 38.17%. After modification under electromotive force gradient of 5 V cm^{-1} and different concentrations of the $CaCl_2$ electrolyte, the mean tensile strength ranged from 1.53 MPa to 2.83 MPa in schemes 3~11. Compared with the tensile strength in scheme 1, the mean tensile strength after the electrochemical treatment increased by 16.79~116.03%.

Scheme	Measured tensile strength (/MPa)						Mean (/MPa)
1	1.22	1.03	1.59	1.13	1.47	1.42	1.31
2	0.81	0.82	0.85	0.74	0.74	0.87	0.81
3	2.27	2.27	2.61	2.33	2.31	2.25	2.34
4	2.09	2.04	2.14	2.11	2.16	2.11	2.11
5	2.80	2.83	2.96	2.78	2.83	2.75	2.83
6	1.61	1.50	2.02	1.77	1.85	1.87	1.77
7	1.51	1.58	1.55	1.50	1.52	1.54	1.53
8	1.94	2.34	2.01	2.22	2.13	1.91	2.09
9	1.99	1.77	1.52	1.72	1.75	1.81	1.76
10	1.68	1.61	1.53	1.58	1.64	1.61	1.61
11	1.66	2.01	1.83	2.11	1.72	1.71	1.84

Table 2. Measured tensile strengths of the samples.

7. Electrochemical modification of the pore structure of mudstone

The combination of micro-CT, digital image processing, and three-dimensional reconstruction is a new, simple, and feasible method for the analysis of the pore structures of mudstone. A single micro-CT image was randomly selected from the 1200 slices of the micro-CT section images. The single digital image was processed by image segmentation, binarization, and compression, and new images were generated with different resolutions. When the pixel size of the new image was taken as the pore aperture, the rule for the variation in rock porosity as the pore aperture varied was estimated from the single micro-CT image. The volume-rendering algorithm of the visualized reconstruction can make the single image the image sequence, and can generate a three-dimensional digital image. The rule for rock porosity variation with variation in the pore aperture was estimated based on the image sequence.

7.1 Three-dimensional reconstruction of micro-CT image sequence

100 micro-CT sections were selected, and the single section processing included image segmentation, binarization, and compression. When processed, the 100 sections were sequenced according to the special algorithm, the micro-CT image sequence and three-dimensional data were generated. Three-dimensional digital image of binary was generated by image preprocessing, three-dimensional reconstruction, and three-dimensional visualization. The rule for the sandstone porosity variation with the pore aperture was estimated.

The compressed algorithm of micro-CT is that: the odd line of the source image is reserved on the X coordinate, the even line of the source image is reserved on the Y coordinate, and then the new matrix composed by the reserved odd and even lines generated compressed image. The pixel size of the new image is increased by 100%.

Because the porosity of micro-CT image is based on the gray scale of the image, the image preprocessing is interpolation and image smoothing, not including gray histogram equalization, image harpening and color process. The distance between the layers of the micro-CT single section is one pixel of 1.94μm, and the distance value is very small. The interpolation algorithm is the gray interpolation, and the image smoothing algorithm is the Gaussian filter.

Ray casting of the volume rendering algorithm was used in the three-dimensional visualization. The two- dimensional projected image was generated through computing the optic effect on all voxels, and the pore structure of sample was shown.

7.2 Electrochemical modification of pore structure in mudstone

After the electromotive force treatment, the three-dimensional digital images of the micro-CT samples in the anodic and cathodic zone were shown in Fig.4. The relationships

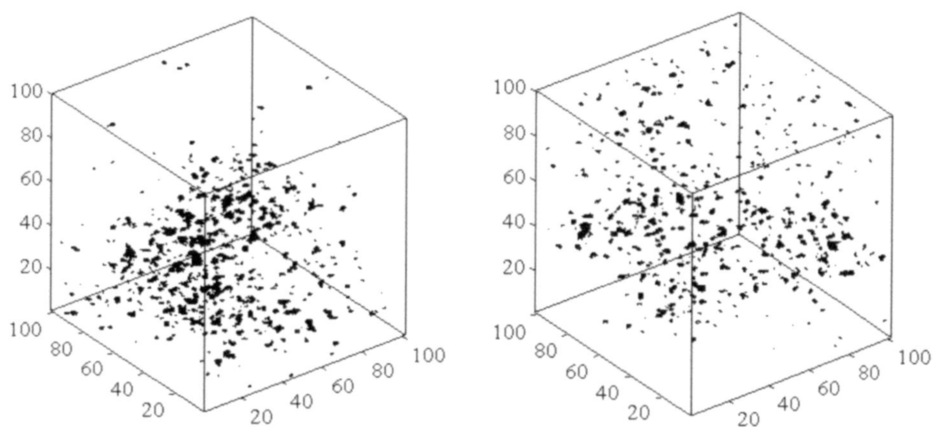

(a) (b)

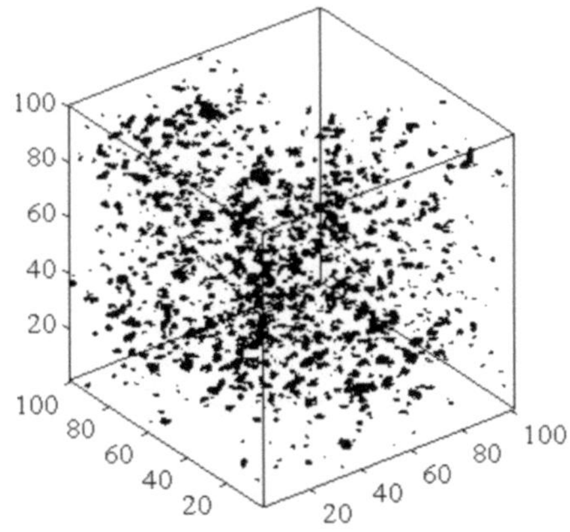

(c)

Fig. 4. Three-dimensional digital images of micro-CT samples: (a) anodic zone; (b) unmodified sample; (c) cathodic zone.

between the variation in sample porosity (n) with variation in pore aperture (P) in the anodic and cathodic zones are shown in Fig. 5. In Fig. 5, the micro-CT image is 2042 × 2042 pixels and the pore aperture is 1.02 μm. After image compression, the new images are 1021 × 1021 pixels, 511 × 511 pixels, and 256 × 256 pixels, and the pore apertures are 2.04 μm, 4.08 μm, and 8.16 μm, respectively. As shown in Fig. 5(a), in the anodic zone, the porosity of the electrochemically modified sample is less than that of the unmodified sample. Before the

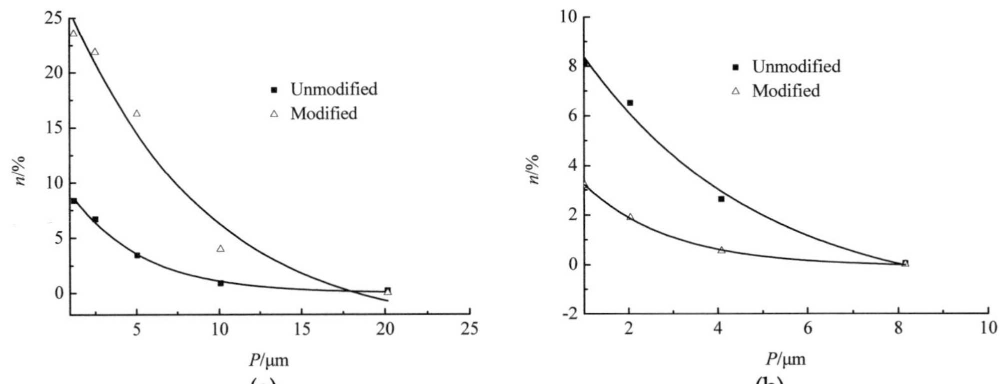

(a) (b)

Fig. 5. The porosity of the sample varies with the pore aperture: (a) anodic zone; (b) cathodic zone.

electrochemical modification, when the pore aperture was 1.02μm, the porosity of the sample was 8.13%. The porosity decreased as the pore aperture increased, and when the pore aperture was then 2.04~8.16 μm, the porosity was 6.51~0.04%. After the electrochemical modification, when the pore aperture was 1.02 μm, the porosity was 3.19%, and when the pore aperture was 2.04~8.16 μm, the porosity was 1.89%~0. When the pore aperture was 1.02 μm, the porosity had decreased by 155% after electrochemical modification. The relationship illustrated in Fig. 5(a) can be expressed as follows:

Unmodified $n=12.99\exp(-0.25P)-1.7$ (12)

Modified $n=5.59\exp(-0.52P)-0.1$ (13)

where P is the pore aperture of the sample and n is the porosity of the sample. The coefficients of correlation are 0.97 and 0.99, respectively.

As shown in Fig. 5(b), the micro-CT image was 2001 × 2001 pixels and the pore aperture was 1.26 μm. The compressed images were 1001 × 1001, 501 × 501, 251 × 251, and 126 × 126 pixels, and the pore apertures were 2.52μm, 5.04μm, 10.08μm, and 20.16μm, respectively. In the cathodic zone, the porosity of the electrochemically modified sample was much higher than that of the unmodified sample. When the pore aperture was 1.26 μm, the porosity of the unmodified sample was 8.37%, and the porosity decreased as the pore aperture increased, so that when the pore aperture was 2.52~20.16μm, the porosity was 6.71~0.2%. After electrochemical modification, the pore aperture was 1.26 μm and the porosity was 23.57%. When the pore aperture was 2.52~20.16 μm, the porosity was 21.84%~0. When the pore aperture was 1.26 μm, the porosity of the electrochemically modified sample was increased by 182%. The relationship illustrated in Fig. 5(b) can be expressed as follows:

Unmodified $n=11.54\exp(-0.23P)-0.03$ (14)

Modified $n=33.24\exp(-0.12P)-3.7$ (15)

The coefficients of correlation were 0.99 and 0.95, respectively.

From equations (12), (13), (14), and (15), it can be seen that as the pore aperture increases, the porosities in the anodic and cathodic zones change according to negative exponential rules. Therefore, the porosities of the samples in both the anodic and cathodic zones are altered by the electromotive force treatment, but the negative exponential relationship between the porosity and the pore aperture cannot be changed.

8. Electrochemical modification mechanism in mudstone

8.1 Changes in the mineralogical composition under electromotive force

Figure 6 shows SEM and X-ray diffraction analyses of mudstone for samples in the anodic and cathodic zones after electromotive force.

(a)

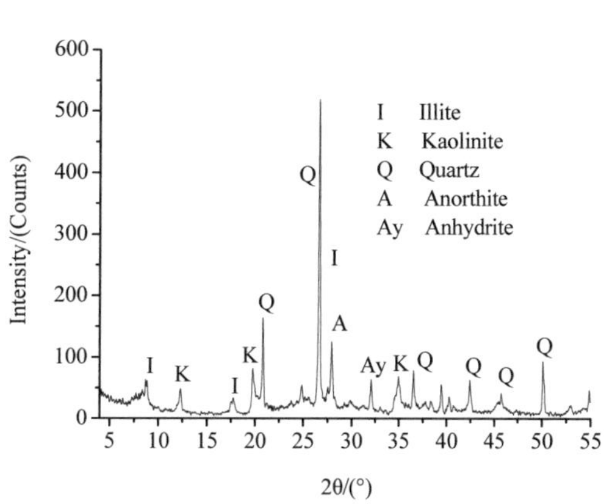

(b)

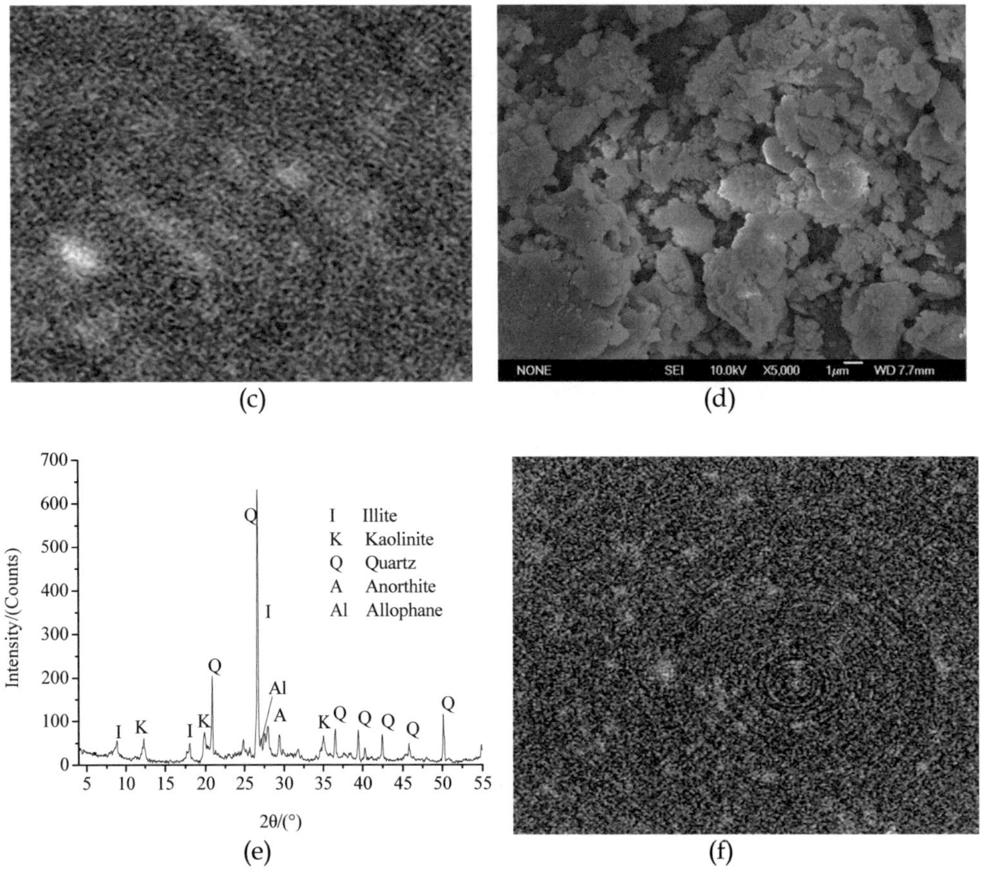

Fig. 6. SEM, X-ray diffraction and micro-CT analyses: (a), (b) and (c) anodic zone; (d), (e) and (f) cathodic zone.

As shown in Fig. 6(a), the mineralogical composition of the anodic-modified sample changed. The new mineral was allophane, and the main mineralogical composition of the sample was illite (38%), kaolinite (8.8%), quartz (30%), anorthite (9%), and allophone (14.2%). As shown in Fig. 6(e), in the cathodic zone, the new mineral was anhydrite and the main mineralogical composition of the sample was illite (40%), kaolinite (7.6%), quartz (30%), anorthite (10%), and anhydrite (12.4%).

Modification of structures and properties with respect to illite, it has a significant change under the action of electromotive force with addition of the electrolyte solutions. X-ray diffraction analyses show that the sheet structure of illite was modified. Modification of structures and properties with respect to kaolinite, it has little significant change. Modification of structures and properties with respect to anorthite, it has been destroyed under electromotive force.

8.2 Analysis of the electrochemical modification mechanism

According to the mineralogical composition of the sample, the main minerals were clay minerals and silicate minerals. Silicate is a semi-conductor, with an electrical conductivity lower than the electrical conductivity of the electrolyte used for this electrochemical treatment. The electrical conductivity of the mudstone mainly depends on the electrical conductivity of the electrolyte in its pores (Jayasekera & Hall, 2007; Revil et al., 2007). The electrochemical modification of the mudstone is mainly affected by the pore structure of the sample and the osmosis and degree of filling of the electrolyte. Figure 7 shows a schematic drawing of the electrochemical modification mechanism based on the pore structure and the mineralogical composition of the mudstone.

As shown in Fig. 7, after the application of a direct current to the mudstone, the electrolyte (E) moves osmotically into the pores and participates in several electrochemical reactions. Electrolysis changes the pH values within the mudstone. Acidification leads to the decomposition or hardening of the silicates and alumina hydroxide, generating allophane from aluminium hydroxide. Alkalization leads to the precipitation of hydroxides on the surfaces of the clay mineral particles, resulting in the generation of a gelatinous precipitate (Pr) and new vug minerals (V). Changes in the pH values cause changes in the mineralogical composition, and new minerals are created, such as allophane and anhydrite, which causes changes in the tensile strength of the mudstone.

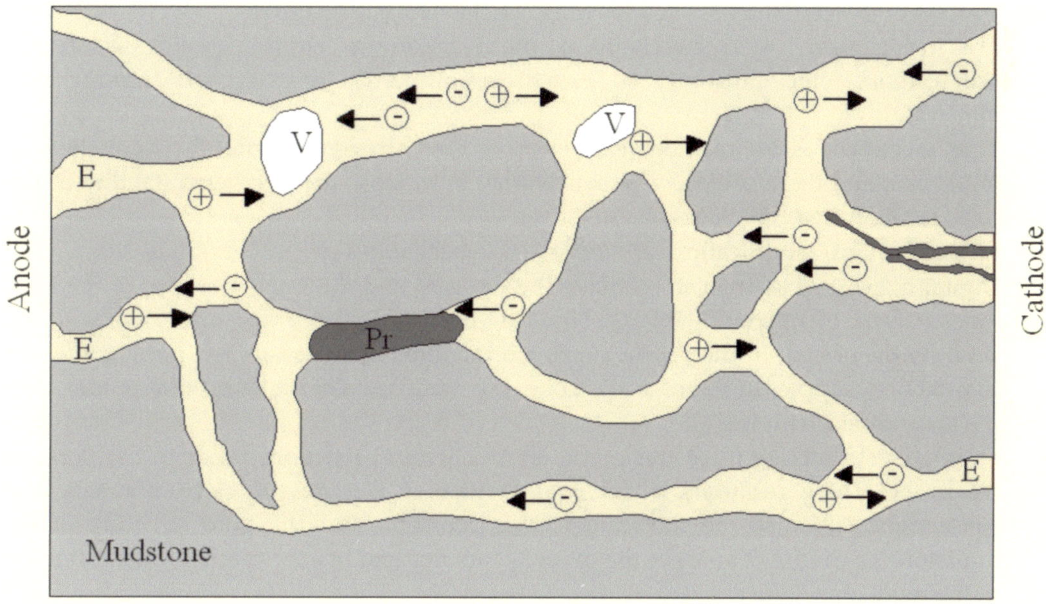

Fig. 7. Schematic drawing of the electrochemical modification of mudstone.

Because of the excess negative charges on the surfaces of the clay mineral particles, the solid particles with negative charges move to the anodic zone and are enriched there by electrophoresis. This causes the fine particles in the pores of the anodic zone to agglomerate and coarsen, so these particles increase in size to fill the pore, causing the porosity to decrease. Electro-osmosis causes the liquid electrolyte, which has a positive charge, to move to the cathodic zone, where it becomes enriched, causing the degree of electro-osmosis to increase. More clay minerals and silicate minerals then take part in the electrolysis, and move toward to the anode. The porosity in the cathodic zone increases.

Electrolysis causes the observed changes in the mineralogical composition of the clay and silicate minerals. Electro-osmosis and electrophoresis cause the observed changes in the pore structure, by causing the positively charged liquid electrolyte to move to the cathode, which in turn causes the hydrated layer to decrease, reduces the hydrophilicity, the dehydration, and the consolidation of the anodic zone, increases the intermolecular and hydrogen bond forces, augments the cohesive and interconnection forces, and increases the cementation properties of the clay mineral particles. As a result, the mechanical strength of the mudstone increases.

9. Conclusion

Under electromotive force, we performed an experiment in which the mudstone in a coal stratum was electrochemically modified. The tensile strengths of the unmodified and modified samples were evaluated with the Brazilian test. The micro-CT experimental system was used for the non-destructive inspection of the samples in the anodic and cathodic zones. The pore structures in the two zones of the samples were analysed with Matlab. The mineralogical compositions of the samples were analysed, and the electrochemical modification mechanism was proposed based on the pore structure and mineralogical composition of the mudstones. The following conclusions were drawn:

- The mechanism of electrochemical modification is electroosmotic dewatering, stabilization, ionic substitutions, structures, properties change, and forming new minerals.
- This electrochemical method can change the physicomechanical parameters of mudstone, and therefore provides a new way to increase the long-term stability of soft rock, facilitating soft rock engineering.
- Electrochemical modification can improve the mechanical properties of mudstone.
- Digital image processing can calculate the porosity and the pore apertures of the rock. The analysis of micro-CT images showed that as the pore aperture increased, the porosity decreased. In the anodic zone, the modified sample was less porous than the unmodified sample. In the cathodic zone, the modified sample was more porous than the unmodified sample.
- During electromotive force treatment, electrochemical reactions occur in the pores of mudstone. These reactions, which mainly involve electrolysis, electrophoresis, and electroosmosis, cause the mineralogical composition and the pore structure of the mudstone to change. These are the main factors that modify the mechanical parameters of the mudstone.

10. Acknowledgment

This research was supported financially by the National Natural Science Foundation of China, grants no. 50474057.

11. References

Adamson, L. G.; Quigley, D. W. & Ainsworth, H. R. (1966). Electrochemical strengthening of clayey sandy soils. *Engineering Geology,* Vol. 1, No.6, pp. 451-459.

Adamson, L. G.; Rieke, I. I. & Grey, R. R. (1967). Electrochemical treatment of highly shrinking soils. *Engineering Geology*, Vol. 2, No.3, (June), pp.197-203.

Adamson, L. G.; Chilingar, G. V. & Beeson, C. M. (1996). Electrokinetic dewatering, consolidation and stabilization of soils. *Engineering Geology*, Vol. 1, No.4, (August), pp. 291-304.

Aggour, M. A. (1992). Muhammadain A M. Investigation of waterflooding under the effect of electrical potential gradient. *Journal of Petroleum Science and Engineering*, Vol. 7, No.3-4, pp. 319-327.

Aggour, M. A.; Tchelepi, H. A. & Al-yousef, H. Y. (1994). Effect of electroosmosis on relative permeabilities of sandstones. *Journal of Petroleum Science and Engineering*, Vol. 11, No.2, (April), pp. 91-102.

Amirat, Y. & Shelukhin, V. (2008). Electroosmosis law via homogenization of electrolyte flow equations in porous media. *Journal of Mathematical Analysis and Applications*, Vol. 342, No.2, (February), pp. 1227-1245.

Chilingar, G. V. (1970). Effect of direct electrical current on permeability of sandstone cores. *Journal of Petroleum Technology*, Vol. 22, No.7, (July), pp. 830-836.

Chukhrov, F. V. (1968). Some Results of the Study of Clay Minerals in the U.S.S.R. *Clays and Clay Minerals*, Vol. 16, (February), pp. 3-14.

Harton, J. H.; Hamid, S. & Abi-Chedid. E. (1967). Effects of electrochemical treatment on selected physical properties of a clayey silt. *Engineering Geology*, Vol. 2, No.3, (June), pp. 191-196.

Jayasekera, S. & Hall, S. (2007). Modification of the properties of salt affected soils using electrochemical treatments. *Geotechnical and Geological Engineering*, Vol. 25, No.1, (January), pp. 1-10.

Revil, A.; Linde, N. & Cerepi, A. (2007). Electrokinetic coupling in unsaturated porous media. *Journal of Colloid and Interface Science*, Vol. 313, No.1, (January), pp. 315-327.

Revil, A. & Jougnot, D. (2008). Diffusion of ions in unsaturated porous materials. *Journal of Colloid and Interface Science*. Vol. 319, No.1, (January), pp. 226-235.

Schlocker, J. & Gray, D. H. (1969). Electrochemical alteration of clay soils. *Clay and Clay Minerals*, Vol. 17, No.2, (April), pp. 309-322.

Sun, H. (2000). *Mechanism and theory study of flowing in porous media under the electric field*. MSD Thesis, Xian, Xian Petroleum Institute, China.

Titkov, N. I. (1961). *Electrochemical induration of weak rocks*, New York, Consultants Bureau, America.

Titkov, N. I.; Petrov, V. P. & Neretina, A. I. (1965). *Mineral formation and structure in the electrochemical induration of weak rocks*. New York, Consultants Bureau, America.

Wang, D.; Kang, T. H. & Chai, Z. Y. (2009). Experimental studies on subsidence and expandability of montmorillonitic soft rock particles under electrochemical treatment. *Chinese Journal Rock Mechanical and Engineer*, Vol. 28, No.9, (September), pp. 1869-1875.

Youell, R. F. (1960). An electrolytic method for producing chlorite-like substances from montmorillonite. *Clay Minerals Bull*, Vol. 9, pp. 43-47.

Zhuang, Y. F.; Wang, Z. & Liu, Q. (2005). Energy level gradient theory for electroosmotic consolidation. *Journal of Institute of Technology*, Vol. 37, No.2, (February), pp. 8-11.

Zhuang, Y. F. & Wang, Z. (2002). Electrokinetic phenomena in soil and their applications. *Joural of Hohai University (natural sciences),* Vol. 30, No.6, (December), pp. 112-115.

4

The EMF Method with Solid-State Electrolyte in the Thermodynamic Investigation of Ternary Copper and Silver Chalcogenides

Mahammad Babanly, Yusif Yusibov and Nizameddin Babanly
Baku State University
Azerbaijan

1. Introduction

Design and optimization of technology of creating new multicomponent inorganic materials, in particular, chalcogenides of metals are perspective functional materials in modern electronic techniques, are based on results of thermodynamic calculations. In turn, to ensure high accuracy of similar calculations is needs reliable dates on fundamental thermodynamic characteristics of corresponding phases.

However, the analysis of literature data shows that unlike binary system, the thermodynamic properties of ternary and multicomponent systems are studied quite insufficiently. In our opinion, elimination of this blank wide application of electromotive forces method (EMF) - one of the most exact experimental methods of chemical thermodynamics appreciably can promote. This method is applied with the big success to the thermodynamic studying of liquid binary and ternary metal systems, which more than 70 % of the available information is concerning to EMF method. Wide application of EMF method for studying of liquid metal systems is caused not only that the specified systems are the most suitable objects of investigation by this method, but also that the mathematical apparatus of chemical thermodynamics allows to calculate strictly integrated thermodynamic functions (ITF) homogeneous binary and ternary system on the basis of values partial thermodynamic functions (PTF) one of components in wide compositions area [M.Babanly et.al, 1992; Morachevskii et.al., 2003; Wagner, 1952].

Unfortunately, specific features and conditions of application EMF method to the heterogeneous systems have not been considered in long time. In the 80th years of last century we have undertaken attempts of elimination of this blank and have developed conditions of application EMF method to heterogeneous metal and semi-conductor systems [M.Babanly, 1985, 1992]. In this chapter we offer the rational method of calculation integral thermodynamic functions of intermediate phases in the ternary heterogeneous systems from PTF of one of components by using of the phase diagram and thermodynamic functions of some boundary binary phases and elementary components. The EMF method with liquid electrolyte has been realized on an example over 30 systems of Thallium-Metal-Chalcogen and have been obtained complexes interconsistency thermodynamic data for the many ternary chalcogenides of thallium [M.Babanly et.al, 1992].

However, it is known, that using a classical variant of EMF method with liquid electrolyte has a number of the restrictions due to with percolation of collateral processes [M.Babanly et.al, 1992; Morachevskii et.al., 2003; Wagner, 1952]. In the works [M.Babanly, 2001, 2009, N.Babanly, 2009, 2010] it is shown, that one of effective ways of expansion of possibilities of EMF method is connected with use solid cationconducting superionic conductors as electrolyte. Advantage of solid electrolytes in comparison with the liquid consists that in them conductivity is carried out by means of ions of one element, and strictly certain charge. It is, firstly, provides a constancy and stability of a charge potential forming ion which is in equilibrium with electrodes of a concentration chains, secondly, solid electrolytes plays a role of the original membrane dividing two electrode spaces and by that prevents many collateral processes due to interaction between electrolyte and electrodes, as well as through electrolyte - between electrodes.

In this chapter is short considered specific features of application of EMF method to heterogeneous systems and results of thermodynamic study of ternary chalcogenides of copper and silver by EMF method with solid electrolytes $Cu_4RbCl_3I_2$ and Ag_4RbI_5 are presented. A part of these thermodynamic data are published earlier, and some are presented for the first time.

2. Some features of application of EMF method to heterogeneous systems

Investigation of thermodynamic properties in homogeneous (liquid or solid state) system A-B usually consists in measurement EMF of concentration chains of type [Wagner, 1952]

$$(-)\ A[sol(liq)]\,|\,A^{z+}\ (\text{in electrolyte})\,|\ A_xB_{1-x}[sol(liq)]\ \ (+) \tag{1}$$

Potential definition process of such elements served reversible electrochemical transfer at constant temperature of a component A from a condition with the big chemical potential (on the left electrode) in a condition with smaller (on the right electrode). Measured EMF (E) at temperature T is directly connected with partial Gibbs energy of a component A in an alloy relative pure element A as standard condition:

$$\Delta\overline{G}_A = \Delta\overline{H}_A - T\Delta\overline{S}_A = -zFE\,, \tag{2}$$

where z–charge of potentialforming ions, F – Faraday number, E – EMF value.

The relative partial molar entropy of component A in A_xB_{1-x} phase composition can be calculate on temperature coefficient of EMF chains (1), as

$$\left(\frac{\partial\Delta\overline{G}_A}{\partial T}\right)_P = -\Delta\overline{S}_A \tag{3}$$

Substitute (2) in (3), we receive:

$$\Delta\overline{S}_A = zF\left(\frac{\partial E}{\partial T}\right)_P \tag{4}$$

The relative partial molar enthalpy can be calculated, substituting the equations (2) and (4) in Gibbs-Helmholtz equation:

$$\Delta\bar{H}_A = \Delta\bar{G} + T\Delta\bar{S}_A$$

$$\Delta\bar{H}_A = -zFE + zFE\left(\frac{\partial E}{\partial T}\right)_p = -zF\left[E + T\left(\frac{\partial T}{\partial T}\right)\right] \quad (5)$$

Thus, measuring equilibrium values of EMF of concentration chains of type (1) in a wide temperature range for various compositions of the right electrodes, it is possible to calculate relative partial molar free energy and entropy of A component in solid solution A_xB_{1-x} at any concentration.

Application of EMF method to heterogeneous systems has some features essentially unlike from homogeneous systems. In the given chapter are shortly considered these features which have allowed to planning experiments correctly at studying of heterogeneous systems what are the overwhelming majority of binary and multicomponent metal and semi-conductor alloys

At investigation of solid-state alloys of binary or more difficult metal and semi-conductor systems by EMF methods should deal with the phase diagrams characterized by various combination homogeneous and heterogeneous areas. In most cases for thermodynamic investigation of phases in such systems, it is necessary to change measurements EMF of chains of type (1) in heterogeneous phase areas. However as shown in [M.Babanly et.al, 1992], despite wide application of EMF method to binary heterogeneous systems, using of EMF measurements results in heterogeneous areas at thermodynamic calculations is not always correctly proved.

If in the homogeneous system A-B the temperature coefficient of EMF of concentration chains of type (1) has physical meaning relative partial molar entropy of A component in alloy, in heterogeneous A-B mix it also reflects change of their compositions with temperature along curves two-phase equilibrium on T-x diagram. By using the results of EMF measurements in two-phase area on T-x diagram is limited to that, unlike relative free energy, relative partial molar entropy and enthalpy discontinuous changes on the border of one- and two-phase areas.

Hence, use of results of EMF measurements of concentration chains in heterogeneous phase areas required the special analysis. Such analysis was carried out in works of [M.Babanly et.al, 1985a, 1992]. We will consider some features of application EMF method to binary heterogeneous systems.

The partial heterogeneous functions (PHF) method offered by G.F.Voronin [Voronin, 1976] for thermodynamic investigation of binary heterogeneous systems, allows to apply to heterogeneous mixes all mathematical apparatus of thermodynamics of solutions, keeping the form of the corresponding thermodynamic equations. Hence, having experimental data on relative partial molar functions of one of components of binary system in all composition field, including homogeneous and heterogeneous phase areas, integration of Gibbs-Duhem equation it is possible to calculate integral thermodynamic functions of separately phases and heterogeneous mixes without any out of thermodynamic assumptions.

Possibility of application of EMF method to binary heterogeneous systems is reduced in definition conditions at which results of EMF measurements of concentration chains of type (1) in heterogeneous areas can be used for calculation partial thermodynamic functions of A component and developing of other calculation procedures of integral thermodynamic functions when direct use of these experimental results is impossible [M.Babanly et.al, 1985a].

Let's see the A-B binary system with the limited solid solutions (α - and β-phases) based on both components. At certain constant temperature T_1 with increasing concentration of B in alloy A-B partial free Gibbs energy of A component continuously decreases in homogeneity area of α-phases, remains constant in two-phase area of α+β $(\Delta\overline{\overline{G}}_A)$ and again continuously decreases in single-phase area of β (fig. 1). According to a condition of chemical equilibrium, at the given temperature

$$(\Delta\overline{\overline{G}}_A) = \Delta\overline{G}_A(x') + \Delta\overline{G}_A(x'') \tag{6}$$

For partial enthalpy and entropy (6) type equality do not follow from the general conditions phase equilibriums. It means rupture of functions $\Delta\overline{H}_A$ and $\Delta\overline{S}_A$ on borders of homogeneity areas of α - and β-phases.

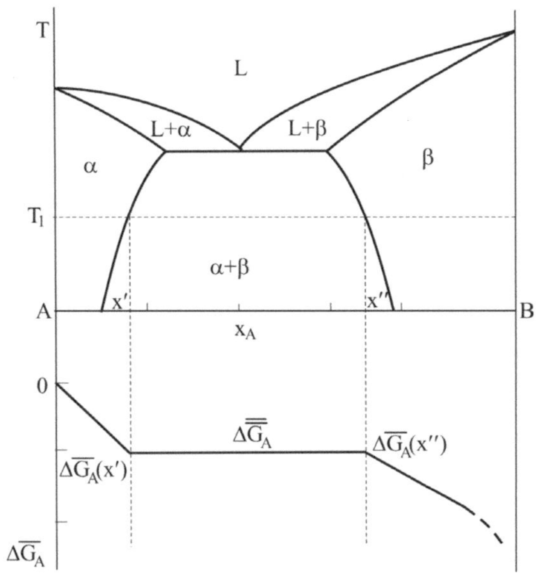

Fig.1. T-x diagram of binary system with limited solid solutions and changes of partial free Gibbs energy with composition.

Compositions of co-existing phases in two-phase area of α+β are function temperature and are defined by the conjugate curves of disintegration (fig. 1). Therefore, differentiating equality (6) on temperature for boundary concentration of α-phases (x'), we will receive

$$\frac{d}{dT}\Delta\overline{\overline{G}}_A = \left(\frac{\partial\Delta\overline{G}_A}{\partial T}\right)_{x'} + \left(\frac{\partial\Delta\overline{G}_A}{\partial x'}\right)_T \left(\frac{dx'}{dT}\right) \tag{7}$$

From this expression taking into account Cibbs-Helmholtz equation

$$\Delta G = \Delta H - T\Delta S \tag{8}$$

follows, that

$$\Delta\bar{\bar{S}}_A = \Delta\bar{S}_A(x') - \left(\frac{\partial\Delta\bar{G}_A}{\partial x'}\right)_T \left(\frac{dx'}{dT}\right) \quad (9)$$

and

$$\Delta\bar{\bar{H}}_A = \Delta H'_A(x') - T\left(\frac{\partial\Delta\bar{G}_A}{\partial x'}\right)_T \left(\frac{dx'}{dT}\right) \quad (10)$$

The similar equations can be received also for boundary concentration of β-phases (x'').

The analysis of the equations (9) and (10) leads to a following important conclusion: at practically vertical borders ($dx'/dT = 0$ и $dx''/dT = 0$) considered two-phase area on T-x to the diagram

$$\Delta\bar{\bar{S}}_A = \Delta\bar{S}_A(x') = \Delta\bar{S}_A(x'') \quad (11)$$

$$\Delta\bar{\bar{H}}_A = \Delta\bar{H}_A(x') = \Delta\bar{H}_A(x'') \quad (12)$$

It is show, that in heterogeneous α+β mixes and co-existing phases the partial entropy and enthalpy of A component are equal between themselves, i.e. on borders between single-phase and two-phase areas discontinuously changes are not observed

EMF values of concentration chains of type (1) and constants of linear equation E=a+bT are connected with PHF equations (2), (4) and (5). Then, for two-phase alloys fairly equality

$$\Delta\bar{\bar{G}}_A = \Delta\bar{G}_A(x') = \Delta\bar{G}_A(x'') = -zFE \quad (13)$$

At observation of conditions (11) and (12) for two-phase alloys the equalities are carried out also

$$\Delta\bar{\bar{S}}_A = \Delta\bar{S}_A(x') = \Delta\bar{S}_A(x'') = zFb \quad 14)$$

$$\Delta\bar{\bar{H}}_A = \Delta\bar{H}_A(x') = \Delta\bar{H}_A(x'') = -zF\left[E + \left(T\frac{\partial E}{\partial T}\right)_P\right] = -zFa \quad (15)$$

At non-observation of conditions (11) and (12) values a and b in $E = a + bT$ equation do not satisfy the equalities (14), (15) and cannot be used for calculation partial thermodynamic functions of A component.

[M.Babanly et.al, 1992] were carried out the similar analysis of behavior partial heterogeneous functions (PHF) in ternary heterogeneous systems and they found, that under condition of vertical position of section borders of phase areas for two-phase equilibrium are fair equalities (11) and (12), and for three-phase area α+β+γ - equalities

$$\Delta\bar{\bar{\bar{G}}}_A = \Delta\bar{G}_A(\alpha') = \Delta\bar{G}_A(\beta') = \Delta\bar{G}_A(\gamma') \quad (16)$$

$$\Delta\bar{\bar{\bar{S}}}_A = \Delta\bar{S}_A(\alpha') = \Delta\bar{S}_A(\beta') = \Delta\bar{S}_A(\gamma') \quad (17)$$

$$\Delta \overline{\overline{H}}_A = \Delta \overline{H}_A(\alpha') = \Delta \overline{H}_A(\beta') = \Delta \overline{H}_A(\gamma') \tag{18}$$

where α', β', γ' - limiting compositions of α-, β- and γ-phases, in the three-phase equilibrium. Thus, at studying of heterogeneous systems by EMF method measurement it is necessary to conduct in intervals of temperatures in which limits of border of section of phase areas are practically vertical. It is a necessary condition for a calculation substantiation partial thermodynamic enthalpy and entropy from EMF measurements in heterogeneous phase areas [M.Babanly et.al, 1985b, 1992]. In the majority of ternary metal and semi-conductor systems in solid state, at temperatures considerably below temperatures melting this condition it is carried out and at correct drawing up is reversible working in the specified temperatures area of an electrochemical chain it is possible to receive experimental data on relative partial molar functions ($\Delta\overline{G}$, $\Delta\overline{H}$, $\Delta\overline{S}$) of one of components.

[M.Babanly et.al, 1992] detailed considered schemes of calculations of integral thermodynamic functions (ITF) from corresponding partial molar value of one of components for various diagram types of a condition ternary systems under condition of vertical position of borders of section of phase areas.

Considering the accepted condition about constancy of coordinates of borders of phase areas independently from temperature in a temperature interval of EMF measurements, various isothermal sections T-x-y diagrams at the specified temperatures ranges should be absolutely identical and quantitatively reflect an arrangement of phase areas on Gibbs triangle. Therefore a basis for an exact choice of limits of integration at calculations is the phase diagram.

Having precisely constructed isothermal section of the phase diagram and the full information about partial molar free energy, entropy and enthalpy for calculation ITF of formation of ternary phases of any compositions, in principle, can be using all methods of thermodynamic calculations applied in thermodynamics of solutions.

Nowadays, for the majority of binary semiconductor phases have reliable data on thermodynamic functions of formation, therefore the most rational way of calculations is integration of Gibbs-Duhem equation on beam sections of type A-ByC1-y (where A - component which used in concentration chain as the left electrode for which are known values relative partial molar thermodynamic functions in homogeneous and heterogeneous areas) [M.Babanly et.al, 1985b, 1992]. Thus the bottom limit of integration, unlike the binary systems, is not one of pure components, two-component alloy of B_yC_{1-y} (homogeneous or two-phase) of boundary system B-C:

$$\Delta Z_{A_x(B_yC_{1-y})_{1-x}} = (1-x)\int_0^x \frac{\Delta\overline{Z}_A}{(1-x)^2}dx + \Delta Z^*_{B_yC_{1-y}} \tag{19}$$

where, $\Delta\overline{Z}_A$-relative partial function of A component on isothermal section of phase diagram, $\Delta Z^*_{B_yC_{1-y}}$ - integral thermodynamic function of formation of homogeneous and heterogeneous alloys B_yC_{1-y}.

The similar approach to calculation of integrated thermodynamic properties of ternary homogeneous systems is offered by Elliott and Chipman [Morachevskii, et.al., 2003].

Advantage of this way of calculation ITF consists is that, all alloys irrespective of their phase composition, are considered as independent investigation objects as in case of a two-component solution, with the partial and integral thermodynamic characteristics. It gives the chance make up integration of Gibbs-Duhem equation in the necessary direction, to be exact, in a direction of composition change in the potentialforming process. In case of binary system this direction from noble component to less noble, and in ternary –from any way chosen B_yC_{1-y} composition of boundary binary system B-C to a A component on corresponding beam section. In other words, possible variants of changes of compositions of various phases in the ternary system in the process of reversible carrying over A in the alloy are formally reduced to composition change on the section A-B_yC_{1-y}, that in the equation (19) variable reflects quantitatively X-mol part of A

In three-phase area composition of co-existing phases and, accordingly, partial molar functions of A component remain constants irrespective of total composition of alloys, and in two-phase (in cases when the direction of beam line A-B_yC_{1-y} does not coincide with connod lines) and single-phase areas is composition functions.

On fig. 2 various types of isothermal phase diagram which reflect all possible variants of alternations of phase areas on the phase diagram are presented:

a. the homogeneity fields of ternary (ABC_2) and binary compounds degenerated in points, and two-phase areas - in direct lines (fig. 2a);
b. on the base of binary compound BC there is a limited homogeneity field (β) on certain section of ternary system A-B-C. It leads to formation of a wide field of two-phase equilibrium (β+C) (fig. 2b); s
c. there are homogeneity field on a base both initial binary (β), and intermediate ternary (γ) compounds (fig. 2c).

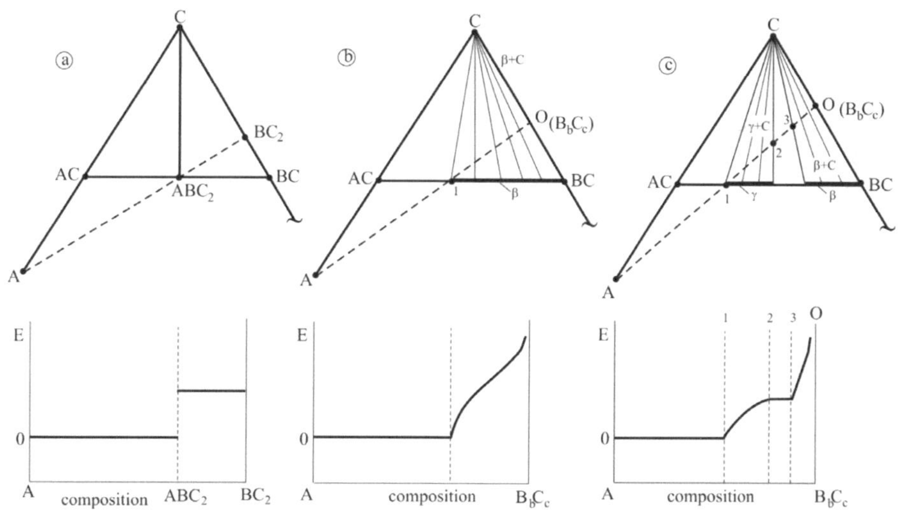

Fig. 2. Some types of phase equilbria in ternary systems and corresponding dependences of E-x on beam sections of type A-B_yC_{1-y} (dot lines on phase diagram) [M.Babanly et.al, 1992]

Apparently from fig. 2, concentration dependences of EMF on beam sections (dot lines) on character coincide with dependences for certain types of diagrams of the phase diagram of binary systems:

a. with binary system with one intermediate compound and imperceptible areas of solid solutions;

b. with binary system with the limited solubility based on more noble initial component;
c. with the binary system, characterized by presence of limited solid solutions based on more noble component and intermediate phase of variable compositions;

Application of the (19) equation to fig. 2 formally differs from binary systems that at x=0, $\Delta Z^{*}_{B_yC_{1-y}} \neq 0$. At calculations value of last function can be borrowed from the literature.

In works of [M.Babanly et.al, 1992] in detail considers calculation procedures ITF for each types of phase given in fig.2

3. Solid-state superion conductors with Ag^{+} and Cu^{+} conductivity

Solid electrolytes - materials possessing high ionic conductivity in a solid state, - are object of comprehensive investigation in various areas of physics and chemistry. From the end of 60th years of the last century interest to this unique class of compounds invariably extended. Solid superionic conductors are the major functional materials of a modern materials science and technology. They are used with the big success as electrochemical sensor controls, electrodes or electrolytes materials in devices of electrochemical transformation energy - in solid-state batteries, displays, high-temperature fuel elements, etc. [Gurevich & Kharkats, 1992; Hagenmuller & Gool,1978; Ivanov-Shits & Murin 2000; West, 1981]. Discovery of solid electrolytes with pure ionic conductivity also has given a new impulse to thermodynamic investigations by electromotive forces method (EMF) and has allowed to extending considerably number of systems studied by this method [M.Babanly et.al, 1992].

Classical example for solid cationconducting electrolytes is high-temperature modification silver iodide α-AgI, existing at temperature above 146°C. High electroconductivity (~1 $Om^{-1}\cdot sm^{-1}$) the given phase which on 4 order exceeds that for low-temperature modification β-AgI have found out in 1914 year by Tubandt and Lorentz. In the range of temperatures from 146 to melting temperature 555°C ionic conductivity of α-AgI monotonously increases, and even in a melting point a little bit decreases [West, 1981].

Silver iodide has appeared good basic compound for synthesis new solid electrolytes possessing high ionic (Ag^{+}) conductivity at room and lower temperatures. It is usually reached by addition to it of the ions which stabilizing cubic structure and interfering its transformation at low temperatures in hexagonal, close-packed on anions. So unipolar solid electrolytes with high ionic conductivity have been synthesized at a room temperature: Ag_8SI, Ag_8SBr, a solid solution $0{,}78AgI\cdot 0{,}22Ag_2SO_4$, $Ag_7I_4PO_4$, $Ag_{19}I_{15}P_2O_7$, $Ag_6I_4WO_4$, and also the gained greatest distribution group of compounds with general formula Ag_4MI_5, where M=Rb, K, NH_4, $Cs_{1/2}$, $K_{1/2}$. Electrolytes with general formula Ag_4MI_5 have one of the highest values of ionic conductivity at a room temperature (~0.2$Om^{-1}\cdot sm^{-1}$) among which it is necessary to allocate so-called "rubidic" electrolyte Ag_4RbI_5 electroconductivity which is long time was record-breaking high (0,25 $Om^{-1}\cdot sm^{-1}$). This compound has superionic conductivity at extremely low temperature - 151°C which there is a phase transition [Ivanov-Shits & Murin, 2000; West, 1981]. At 64°C Ag_4RbI_5 undergoes second sort phase transition, however for this reason electroconductivity changes continuously.

The Ag_4RbI_5 melts at 503K with decompose on peritectic reaction, and below 300 K it decomposes on solid state reaction [Ivanov-Shits & Murin, 2000]. However, last process kinetically is strongly broken and at observance of certain care (absence of a moisture and iodine vapor) Ag_4RbI_5 can be cooled easily without decomposition below a room temperature and to use as solid electrolyte.

In the beginning of 70th years of lost century have been synthesized the Cu^+ conducting superionic conductors which mainly halides of copper. In 1979 the Japanese and independently American chemists have been synthesized related "rubidic" electrolyte the solid electrolyte $Cu_4RbCl_3I_2$ possessing at a room temperature record-breaking high (0,5 $Om^{-1} \cdot sm^{-1}$) ionic conductivity on Cu^+ cations [Gurevich & Kharkats 1992; Ivanov-Shits & Murin, 2000].
The discovery of solid-state electrolytes with pure Cu^+ and Ag^+ conductivity was stimulate to thermodynamic investigation of systems based on copper and silver by EMF method.
The electrochemical cell in the EMF method with solid cationconducting electrolyte like:

$$(-)\ \text{A (solid or liquid)}\ |\ \text{ionic conductor on } A^{z+}\ |\ \text{A in alloy (solid or liquid)}\ (+) \quad (20)$$

Where the left electrode has pure A component, and right - a homogeneous or heterogeneous alloy of multicomponent system.
As solid electrolytes, basically, superionic conductors with pure ionic conductivity are used, as only in this case there is unique dependence between EMF value (E) and Gibbs energy of potentialforming reactions, in condition of constants charge of the ion causing electroconductivity. Presence of electronic making conductivity leads to decrease of EMF value of cells in comparison with its thermodynamic value as active electrons cause short circuit electrolytic chains through internal resistance of electrolyte. As a result of it in cells with the mixed conductors is not reached the equilibrium condition [M.Babanly et.al, 1992].
The solid electrolytes divides two electrode spaces and the last can contain solid phases, liquids or gaseous substances of the identical or various chemical natures. For example, on both side of solid electrolyte there can be gaseous oxygen at two kinds of various partial pressures.
For the first time the electrochemical cell of type (20) has been used for thermodynamic research of two-component system of Ag-S, where as solid electrolyte served β-AgI [West, 1981]. For this purpose it was measured of EMF of concentration chain

$$(-)\ \text{Ag (solid)}\ |\ \beta\text{-AgI (solid)}\ |\ Ag_2S\ \text{(solid)+S (liquid), graphite}\ (+)$$

at the 450-550 K temperature range.
Considering, that in system Ag-S is formed only one binary compound Ag_2S with narrow homogeneity area, and solubility of Ag_2S in liquid sulphur in the specified temperatures range is insignificant, potentialforming reaction can be given as:

$$\text{Ag (solid)+0,5 S (liquid)=0,5 } Ag_2S \text{ (solid)}$$

From experiments have received temperature dependence of EMF and authors have calculated, $\Delta G°$, $\Delta H°$ and $\Delta S°$ of Ag_2S

4. Thermodynamic investigation of the ternary chalcogenides based on copper and silver by EMF method with solid-state electrolytes

Chalcogenides of copper and silver with p^1-p^3 elements [Max Plank Institute, 1992-1995; M.Babanly et.al, 1993; Shevelkov, 2008] have the practical interest as functional materials of electronic techniques (photoelectric, thermoelectric and magnetic semiconductors, superconductors, superionic conductors etc.).
Phase equilibriums in the specified systems are studied in the many works which are results are periodically systematized and critically analyzed in a number of handbooks and monographies [Max Plank Institute, Stuttgart, 1992-1995; M.Babanly et.al, 1993].

For investigation solid-phase equilibria in the systems A-B-X (A-Cu, Ag; B-elements of subgroups of gallium, germanium, arsenic; X-S, Se, Te) and thermodynamic properties of ternary compounds formed in them we had been made concentration chains of types:

$$(-)\ Cu\ (solid)\ |\ Cu_4RbCl_3I_2\ (solid)\ |\ (Cu\ in\ alloy)\ (solid)\ (+) \tag{21}$$

$$(-)\ Ag\ (solid)\ |\ Ag_4RbI_5(solid)\ |\ (Ag\ in\ alloy)\ (solid)\ (+) \tag{22}$$

The equilibrium alloys from various phase areas of the above-mentioned systems served as the right electrodes.

The compound $Cu_4RbCl_3I_2$ synthesized by melting stochiometric amounts of chemically pure, anhydrous CuCl, CuI and RbCl in evacuated (~10^{-2} Pa) quartz ampoule at 900 K with the cooling to 450K and annealed at this temperature for 100 h. Ag_4RbI_5 synthesized from chemically pure RbI and AgI by a technique [West, 1981]: stochiometric mix initial iodides have co-melted in evacuated quartz ampoule (~10^{-2}Pa) and then quickly cooled to a room temperature. At cooling melt crystallizes in fine-grained and microscopic homogeneous state and then annealed at 400K for 200 h. Obtained cylindrical ingots in diameter ~8mm cuts like pellets in the thickness of 4-6 mm which used as solid electrolyte in chains of types (21) and (22).

The elementary copper and silver served as left electrodes and the right electrodes –pre-synthesized and annealed alloys of investigated systems from various phase areas. Compositions of alloys have been chosen from data on phase equilibrium. For preparation of the right electrodes annealed alloys grinded as powder, and then pressed as pellet in weight of 0,5-1 gram. The electrochemical cell of type in fig. 3 has filled with argon and has placed in the tube furnace, where it held at ~380K for 40-50 hours. Cell temperature measured by chromel-alumel thermocouples and mercury thermometers with accuracy±0,5^{0}C.

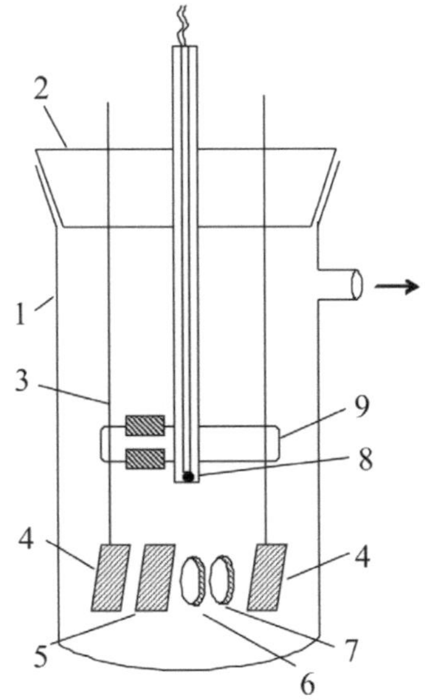

Fig. 3. The electrochemical cell for EMF measurement of chains of type (4.1) and (4.2). 1-glass block; 2-cover; 3-platinium wires; 4-platinium plates; 5-copper (silver) plate; 6-solid electrolytes; 7-investigated alloy (the right electrode); 8-thermocouple, 9-clip [M.Babanly, et.al. 2009].

EMF measured by the compensation method by means of high-resistance digital voltmeters B7-34A. Measurements were carried out in each 3 hours after an establishment of certain temperature. Equilibrium considered those values of EMF which at repeatedly measurement at the given temperature differed from each other not more than on 0,5 mV irrespective from direction of temperature change. In order to of occurrence elimination thermo-e.m.f. contacts of all leads with copper wires had identical temperature.
EMF measurements of alloys of selenium and tellurium containing systems are carried out in the range temperatures of 300÷420K, and sulphur containing 300÷380 K. Maximum limits of temperature intervals of EMF measurements are chosen to exclude melting and transition in a metastable state of alloys of the right electrodes.
Processing of the EMF measurements results. For the thermodynamic calculations the results of the experiments were used, which are satisfying to criteria of the reversible work of a chain. Results of EMF measurements for alloys with different compositions within one heterogeneous area were processed in common.
Measured equilibrium values of EMF put on E=f (T) diagram. Appreciable deviations from linear dependence of EMF were not observed. It is indirectly confirms a constancy of compositions of existing phases in heterogeneous areas of the investigated systems, that, as shown above, is a necessary condition for carrying out of thermodynamic calculations according to the EMF measurements of chains of type (21) and (22). Considering this, results of EMF measurements processed by the least squares method [Gordon, 1976]. Temperature dependence is expressed by the linear equation

$$\mathrm{E} = \mathrm{a} + b\mathrm{T} \equiv \overline{\mathrm{E}} + b(T - \overline{T}) \; . \tag{23}$$

Here $\overline{\mathrm{E}} = \frac{\sum \mathrm{E_i}}{n}$, $\overline{\mathrm{T}} = \frac{\sum \mathrm{T_i}}{n}$, $b = \frac{\sum (\mathrm{E_i}\text{-}\mathrm{E})(\mathrm{T_i} - \overline{\mathrm{T}})}{\sum \mathrm{E_i}(\mathrm{T_i} - \overline{\mathrm{T}})^2}$, where, E_i–experimental values of EMF at temperature T_i; n – number of experimental points (both values E and T), $a = \overline{E} - b\overline{T}$.
The statistical estimation of error of measurements consisted in calculation of dispersions of individual measurements of EMF (δ_E), average EMF values (δ_{E}^2), and also coefficients a (δ_{a}^2) and b (δ_{b}^2) on relations

$$\delta_E(\mathrm{T}) = \frac{\delta_E^2}{n} + \delta_{\mathrm{b}}^2(\mathrm{T} - \overline{\mathrm{T}})^2$$

$$\delta_E^2 = \frac{\sum (\mathrm{E_i} - \tilde{\mathrm{E}}_\mathrm{i})^2}{n-2}$$

$$\delta_{\mathrm{a}}^2(\mathrm{T}) = \frac{\delta_E^2}{n} + \frac{\delta_{\mathrm{E}}^2 \overline{T}^2}{\sum (\mathrm{T_i} - \overline{\mathrm{T}})^2}$$

$$\delta_{\mathrm{B}}^2(\mathrm{T}) = \frac{\delta_{\mathrm{E}}^2}{\sum (\mathrm{T_i} - \overline{\mathrm{T}})^2}$$

$\tilde{\mathrm{E}}_\mathrm{i}$ - EMF values, calculated by (23) equation at temperature T_i. Errors (Δ_i) corresponding values calculated by the relations

$$\Delta_i = t\delta_i$$

(t–Student's test, δ_i –standard deflection). In the present work n≥20, that at confidential level of 95 % leads to t≈2 [Gordon, 1976].
The accepted equations of temperature dependences of EMF according to the recommendation of [Kornilov et.al, 1972] are presented as:

$$E = a + bT \pm 2\left[\frac{\delta_E^2}{n} + \frac{\delta_E^2(T-\bar{T})^2}{\sum(T_i-\bar{T})^2}\right]^{\frac{1}{2}} \quad (24)$$

From the accepted equations of type on relations (13) - (15) calculated relative partial molar free Gibbs energy, enthalpy and entropy of copper (silver) in alloys at 298K.
The Cu-Tl-Te system is studied by EMF measurement of concentration chains of type (21) in the Tl_2Te-Cu_2Te-Te composition field and with taking into account literature data [Max Plank Institute, Stuttgart, 1992-1995; M.Babanly, 1993] the fragment of the solid phase equilibrium diagram (fig. 4) is constructed.

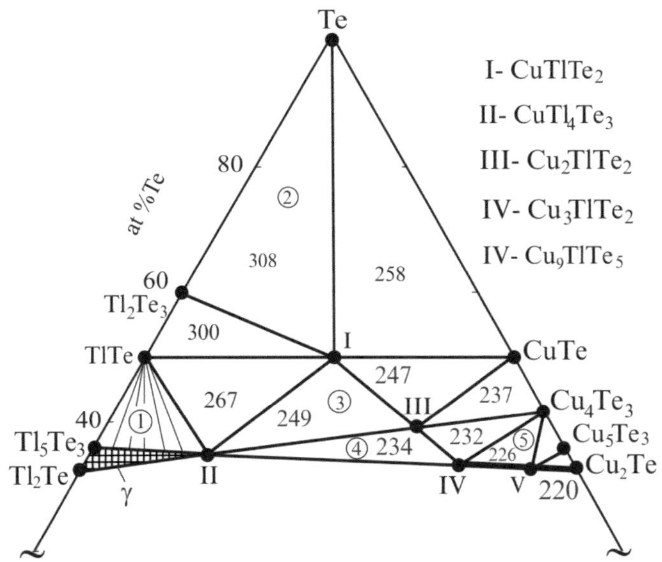

Fig. 4. The solid phase equilibrium diagram of the Cu-Tl-Te system. In some solid phase fields are given the EMF values (mV) of chains of type (5.1) at 300K.

In fig.4 we can see, that in the specified compositions fields five ternary compounds are formed. Compound $CuTl_4Te_3$ forms continuous solid solutions (γ) with Tl_5Te_3. Areas of homogeneity of other ternary and binary compounds of the system are insignificant.
The EMF measurements of chains of type (21) have shown that values of electromotive forces in each three-phase areas on fig. 4 are constant irrespective of total composition of alloys and in discontinuous change on their borders, and in two-phase area TlTe+γ and within homogeneity area γ-phases continuously change depending on composition of last.
Reproducibility of EMF measurements and conformity of sequence of their change in investigated system the thermodynamic conditions (impossibility of reducing of EMF values

in direction $Cu \rightarrow Tl_xTe_{1-x}$) to specify possibility of use of these data for thermodynamic calculations.
For calculation of thermodynamic functions of ternary compounds and γ-phases of variable composition have been used data of measurements in phase areas №№ I-V on fig. 4 (table 1).

№	Phase area	$E, mV = a + bT \pm t \cdot S_E(T)$
1	$TlTe+\delta(Cu_{0,2}Tl_{4,8}Te_3)$	$317,4 + 0,138T \pm 2\left[\frac{0,7}{26} + 3,8 \cdot 10^{-5}(T - 351,2)^2\right]^{1/2}$
2	$TlTe+\delta(Cu_{0,4}Tl_{4,6}Te_3)$	$296,8 + 0,086T \pm 2\left[\frac{0,6}{26} + 3,6 \cdot 10^{-5}(T - 351,2)^2\right]^{1/2}$
3	$TlTe+\delta(Cu_{0,6}Tl_{4,4}Te_3)$	$290,1 + 0,032T \pm 2\left[\frac{1,3}{26} + 9,8 \cdot 10^{-5}(T - 351,2)^2\right]^{1/2}$
4	$TlTe+\delta(Cu_{0,8}Tl_{4,2}Te_3)$	$279,3 + 0,009T \pm 2\left[\frac{1,8}{26} + 1,2 \cdot 10^{-4}(T - 351,2)^2\right]^{1/2}$
5	$\delta(CuTl_4Te_3)$	$281,7 - 0,047T \pm 2\left[\frac{1,5}{26} + 1,1 \cdot 10^{-4}(T - 351,2)^2\right]^{1/2}$
6	$Tl_2Te_3+CuTlTe_2+Te$	$286,3 + 0,073T \pm 2\left[\frac{3,2}{24} + 1,5 \cdot 10^{-4}(T - 353,7)^2\right]^{1/2}$
7	$CuTlTe_2+Cu_2TlTe_2$	$224,3 + 0,085T \pm 2\left[\frac{2,1}{24} + 1,2 \cdot 10^{-4}(T - 353,7)^2\right]^{1/2}$
8	$Cu_2TlTe_2+Cu_3TlTe_2$	$216,5 + 0,066T \pm 2\left[\frac{1,9}{24} + 8,9 \cdot 10^{-5}(T - 353,7)^2\right]^{1/2}$
9	$Cu_3TlTe_2+Cu_9TlTe_5+Cu_4Te_3$	$200,4 + 0,091T \pm 2\left[\frac{2,6}{24} + 1,4 \cdot 10^{-4}(T - 353,7)^2\right]^{1/2}$

Table 1. The temperature dependences of EMF of concentration chains of type (21) in some phase areas of the Cu-Tl-Te system (T=300÷420K).

From the accepted equations of temperature dependences of EMF (tab. 1) the relative partial thermodynamic functions of copper in alloys at 298K (tab. 2) are calculated.
Isotherms of partial thermodynamic functions of copper on the section Tl_5Te_3-$CuTl_4Te_3$ (fig. 5) have continuous curves that specifies in formation of a continuous number of solid solutions between these compounds.

Phase area	$-\overline{\Delta G}_{Cu}$	$-\overline{\Delta H}_{Cu}$	$\overline{\Delta S}_{Cu}$,
	kJ·mole-1		J·K-1·mole-1
Tl_2Te_3+$CuTlTe_2$+Te	29,723±0,142	27,62±084	7,04±2,36
$CuTlTe_2$+Cu_2TlTe_2	24,086±0,123	21,64±0,74	8,20±2,11
Cu_2TlTe_2+ Cu_3TlTe_2	22,787±0,104	20,89±0,65	6,37±1,82
Cu_3TlTe_2+Cu_9TlTe_5+Cu_4Te_3	21,953,0,136	19,34±0,82	8,78±2,28
TlTe+δ($Cu_{0,2}Tl_{4,8}Te_3$)	34,593±0,071	30,59±0,42	13,32±1,19
TlTe+δ($Cu_{0,4}Tl_{4,6}Te_3$)	31,110±0,070	28,64±0,41	8,30±1,16
TlTe+δ($Cu_{0,6}Tl_{4,4}Te_3$)	28,911±0,110	27,99±0,67	3,09±1,91
TlTe+δ($Cu_{0,8}Tl_{4,2}Te_3$)	27,208±0,123	26,95±0,74	0,86±2,11
δ($CuTl_4Te_3$)	25,829±0,117	27,18±0,69	-4,53±2,02

Table 2. The relative partial thermodynamic functions of copper in Cu-Tl-Te alloys at 298 K.

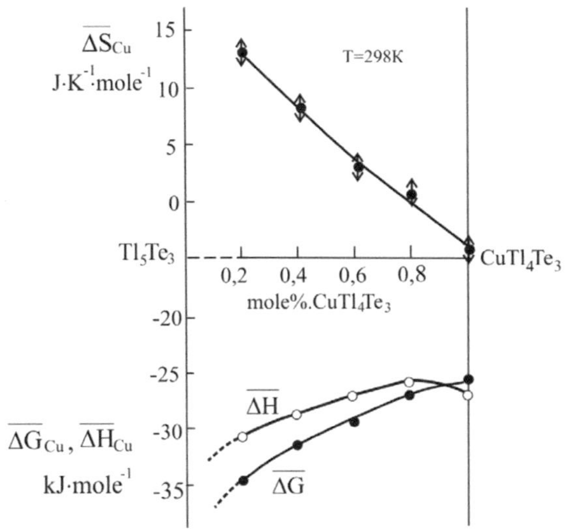

Fig. 5. Dependences of partial thermodynamic functions of copper with compositions on the Tl_5Te_3-$CuTl_4Te_3$ section at 298K.

According to fig. 5, increasing of concentration of copper in solid solutions is accompanied by considerable decreasing of composition-sensitive function of $\overline{\Delta S}_{Cu}$ that good agree with structural data of $CuTl_4Te_3$ and γ-phases [M.Babanly et.al, 1993].

The standard thermodynamic function of formation of γ-phase and ternary compound $CuTl_4Te_3$ are calculated by graphical integration of Gibbs-Duhem equation on the beam section of Cu-[Tl_4Te_3] (where, [Tl_4Te_3]-two-phase mix Tl_5Te_3 и TlTe).

Insignificance of homogeneity areas of ternary compounds $CuTlTe_2$, Cu_2TlTe_2, Cu_3TlTe_2 and Cu_9TlTe_5 and co-existing phases with them (Tl_5Te_3, Cu_4Te_3 and Te) in three-phase areas №№ II-V (fig. 4) allows calculate their standard thermodynamic functions of formation and standard entropy by method potentialforming reactions. According to fig. 4, the partial molar functions of copper in the specified phase areas are thermodynamic characteristics of following potentialforming reactions (all compounds in crystalline state):

$$Cu+0{,}5Tl_2Te_{3+0{,}}5Te=CuTlTe_2 \quad (4.5)$$

$$Cu+CuTlTe_2=Cu_2TlTe_2 \quad (4.6)$$

$$Cu+Cu_2TlTe_2 =Cu_3TlTe_2 \quad (4.7)$$

$$Cu+0{,}5Cu_3TlTe_2+0{,}5Cu_4Te_3=0{,}5Cu_9TlTe_5 \quad (4.8)$$

Based on these reactions the thermodynamic functions of formation and standard entropy of corresponding ternary phases have been calculated.
At calculations besides experimental data (tab. 2) the corresponding thermodynamic data for compounds Tl_2Te_3, TlTe, Tl_5Te_3 [M.Babanly et.al, 1993.] and Cu_4Te_3 [Abbasov, 1981], also standard entropy of copper and tellurium [Yungman, 2006] were used. Errors calculated a method of accumulation of errors.
The obtained values of standard thermodynamic functions of formation and standard entropy of ternary compounds Cu_2TlTe_2, Cu_3TlTe_2 and Cu_9TlTe_5 are well agree with the results [M.Babanly et.al, 2011] which is obtained from EMF measurements of concentration chains concerning a thallium electrode with liquid electrolit (tab. 3).

Compound	$-\Delta_f G^0(298K)$	$-\Delta_f H^0(298K)$	$S^0(298K)$,
	kJ·mole^{-1}		J·K^{-1}·mole^{-1}
Tl_5Te_3 [31]	213,6±1,7	216,7±2,0	458,6±6,7
$\delta(Cu_{0,2}Tl_{4,8}Te_3)$	210,2±1,7	213,0±2,2	454±7
$\delta(Cu_{0,4}Tl_{4,6}Te_3)$	207,8±1,6	210,5±2,3	449±7
$\delta(Cu_{0,6}Tl_{4,4}Te_3)$	205,3±1,6	207,6±2,4	444±8
$\delta(Cu_{0,8}Tl_{4,2}Te_3)$	203,8±1,5	206,0±2,5	438±8
$CuTl_4Te_3$	201,4±1,4	203,8±2,6	433±9

Table 3. The standard integral thermodynamic functions of solid solutions $Cu_xTl_{5-x}Te_3$ $(0<x<1)$.

From EMF measurements of concentration chains of types (21) and (22) the standard thermodynamic functions of formation and standard entropy of some ternary chalcogenides of copper (tab. 4) and silver (tab. 5) are calculated. Herein, some data are published for the first time, which are the sources do not show. In the tab.4 and 5 also presence the thermodynamic functions of thallium containing ternary compounds of copper and the silver which are obtained by EMF method with liquid electrolyte (Italic font). The data presented in tables strongly differ on errors. It is due to with different errors of thermodynamic functions of the binary compounds which are recommended in the handbooks [M.Babanly et.al, 1992; Kubaschewski, 1993; Mills, 1974, Yungman, 2006] which have been used at calculations.

Compound	$-\Delta_f G^0(298)$	$-\Delta_f H^0(298)$	$S^0(298)$	References
	kJ/mole		$J \cdot K^{-1} \cdot mole^{-1}$	
$CuIn_3Se_5$	380,0±1,4	398,2±28,6	373±28	[Babanly, N.B., 2009]
$CuInSe_2$	153,2±0,6	158,0±9,6	163±11	[Babanly, N.B., 2009]
$CuTlS_2$	94,3±0,7 *91,5±0,5*	93,6±1,4 *98,6±4,0*	172,7±2,8	[Babanly, N.B., 2009] [Babanly, M.B., 1986]
$CuTlS$	90,3±0,7 *84,1±1,5*	88,3±2,1 *82,1±4,9*	132,4±6,2	[Babanly, N.B., 2009] [Babanly, M.B., 1986]
Cu_3TlS_2	163,8±2,6 *152,7±1,8*	159,2±9,8 *145,8±12,3*	251,8±5,8	[Babanly, N.B., 2009] [Babanly, M.B., 1986]
Cu_9TlS_5	373,8±3,9 *354,6±4,5*	371,8±21,4 *339,7±30,8*	529,0±19,0	[Babanly, N.B., 2009] [Babanly, M.B., 1986]
$CuTlSe_2$	96,29±0,16 *96,5±0,6*	97,91±0,95 *97,2±1,3*	176,1±5,1	[Babanly, M.B., 1992]
$CuTlSe$	84,49±0,16 *84,2±1,3*	81,37±0,85 *80,5±3,9*	149,9±2,8	[Babanly, M.B., 1992]
Cu_2TlSe_2	119,06±0,27	118,61±1,54	216,2±6,8	
$CuTlTe_2$	75,1±0,4	72,6±1,3	208±4	
Cu_2TlTe_2	99,2±0,5 *94,8±0,9*	94,3±2,1 *92±7*	249±6 *237±3*	[Babanly, M.B., 2010]
Cu_3TlTe_2	122,0±0,6 *117,1±1,2*	115,2±2,7 *117±5*	288±8 *263±4*	[Babanly, M.B., 2010]
Cu_9TlTe_5	264,3±2,6 *244,0±2,4*	253,8±9,8 *2431±14*	637±15 *621±7*	[Babanly, M.B., 2010]
$CuTl_4Te_3$	201,4±1,4	203,8±2,6	433±9	
Cu_2GeS_3	225±12	226±13	189±8	[Babanly, M.B. 2001]
Cu_8GeS_6	456±13	429±18	579±24	[Babanly, M.B. 2001]
Cu_2GeS_6	182±11	182±11	35	[Babanly, M.B. 2001]
$CuGe_3Se_4$	252±56	246±55	314,5±8,4	[Babanly, M.B. 2001]
$CuGeSe_2$	131±21	129±21	154,9±7,0	[Babanly, M.B. 2001]
Cu_2GeSe_3	178,4±18,8	174,5±19,7	223,4±6,6	[Babanly, M.B. 2001]
Cu_2GeTe_3	89,6±13	92±14	236±12	[Babanly, M.B. 2001]
$Cu_2Sn_4S_9$	659,9±4,3	650,9±29,7	560,3±74,7	[Babanly, M.B. 2001]
Cu_2SnS_3	239,6±1,5	242,6±12,0	196,3±21,9	[Babanly, M.B. 2001]
Cu_4SnS_4	316,4±2,4	327,7±18,8	266,5±28,2	[Babanly, M.B. 2001]
Cu_2SnSe_3	205,5±2,2	207,6±14	244±27,9	[Babanly, M.B. 2001]
Cu_2SnTe_3	124,3±1,9	118,7±11,6	278±15,9	[Babanly, M.B. 2001]
Cu_3AsS_4	179,2±0,6	172,2±2,6	278±8	
$Cu_6As_4S_9$	429,4±1,2	419,5±8,2	673±23	
$Cu_4As_2S_5$	257,8±0,8	249,8±4,6	395±13	
Cu_3AsS_3	170,2±0,6	163,9±2,7	254±8	
$CuAsS$	69,5±0,3	64,1±1,7	109±5	
$Cu_2As_4Se_7$	206,7±1,6	208,1±10,6	501±17	

Compound	$-\Delta_f G^0(298)$	$-\Delta_f H^0(298)$	$S^0(298)$	References
	kJ/mole		$J \cdot K^{-1} \cdot mole^{-1}$	
$CuAsSe_2$	74,6±0,5	73,8±3,2	156±6	
Cu_3AsSe_4	163,5±1,1	157,2±5,6	325±12	
$Cu_4As_2Se_5$	233,6±2,4	227,9±8,8	472±21	
Cu_3AsSe_3	158,0±1,3	151,5±5,3	296±13	
Cu_3AsTe_3	91,2±3,3	90,6±5,3	286±6	[Babanly, M.B. 2001]
Cu_3SbS_4	207,5±3,9	200±6,3	298±18	[Babanly, M.B. 2001]
$CuSbS_2$	121,4±3,4	119±3,5	148±6,4	[Babanly, M.B. 2001]
Cu_3SbS_3	197,8±3,8	189,3±6,5	269,5±13,7	[Babanly, M.B. 2001]
Cu_3SbSe_4	191,6±2,5	178,6±5,4	358±18	[Babanly, M.B. 2001]
$CuSbSe_2$	101,4±1,8	98,5±2,2	173±8	[Babanly, M.B. 2001]
Cu_3SbSe_3	175,6±2,5	164,0±5,3	311±15	[Babanly, M.B. 2001]
$CuBiS_2$	138,6±4,0	138,2±2,9	156±12	[Babanly, M.B. 2001]
Cu_3BiS_3	213,0±4,4	209,9±5,2	264±21	[Babanly, M.B. 2001]
$CuBi_3Se_5$	248,7±1,9	248,6±5,8	421,9±7,8	[Babanly, M.B. 2001]
$CuBiSe_2$	107,6±0,8	105,9±2,51	189,8±2,4	[Babanly, M.B. 2010]
Cu_3BiSe_3	162,5±1,2	155,9±5,7	315,0±8,5	[Babanly, M.B. 2010]
Cu_9BiSe_6	324,8±3,5	313,1±18,6	659±28	[Babanly, M.B. 2010]
$CuBiTe_2$	64,2±1,0	61,3±1,0	200±7	[Babanly, N.B. 2007]

Table 4. Standard thermodynamic functions of formation and standard entropy of some ternary chalcogenides of copper.

Compound	$-\Delta_f G^0(298)$	$-\Delta_f H^0(298)$	$S^0(298)$	References
	kJ/mole		$J \cdot K^{-1} \cdot mole^{-1}$	
$AgGaS_2$	302,1±1,7	302,8±4,3	145,1±9,6	[Ibragimova G.I., 2006]
Ag_9GaS_6	447,5±2,4	393,9±12,4	786,8±27,8	[Ibragimova G.I., 2006]
$Ag_2Ga_{20}S_{31}$	5131±21	5169±62	1772±87	[Ibragimova G.I., 2006]
$AgGaSe_2$	237,0±3,4	239,4±5,6	159,6±11,2	[Ibragimova G.I., 2006]
Ag_9GaSe_6	433,0±4,1	413,1±10.9	742,9±32,5	[Ibragimova G.I., 2006]
$AgGaTe_2$	120,8±4,6	119,6±3,1	186,8±6,9	[Ibragimova G.I., 2006]
Ag_9GaTe_6	276,8±11,5	233,4±11,0	867,0±30,7	[Ibragimova G.I., 2006]
Ag_2GeS_3	206±2,1	198±2,2	239,1±8,8	[Babanly, M.B., 1993]
Ag_4GeS_4	254±2,1	235±2,4	393,2±14,1	[Babanly, M.B., 1993]
Ag_8GeS_6	345±2,2	310±2,6	680,4±23,1	[Babanly, M.B., 1993]
Ag_2GeSe_3	145±2,1	139±2,2	262,2±10,4	[Babanly, M.B., 1993]
Ag_8GeSe_6	288±2,3	255±2,8	734,6±30,4	[Babanly, M.B., 1993]
Ag_8GeTe_6	268,0±1,0	245,0±7,0	745,8±17,1	[Babanly, M.B., 1993]
$Ag_2Sn_2S_5$	358,8±2,3	339,5±12,6	397,9±16,3	[Babanly, M.B., 1993]
Ag_2SnS_3	213,3±1,6	202,7±8,8	260,7±16,8	[Babanly, M.B., 1993]
Ag_8SnS_6	351,7±2,6	328,9±18,0	652,9±16,3	[Babanly, M.B., 1993]
Ag_8SnS_5	355±3,2	330±3,8	628,43±23,6	[Babanly, M.B., 1993]

Compound	$-\Delta_f G^0(298)$	$-\Delta_f H^0(298)$	$S^0(298)$	References
	kJ/mole		$J{\cdot}K^{-1}{\cdot}mole^{-1}$	
$AgSnSe_2$	146,4±0,5	148±3	162±7,2	[Babanly, M.B., 1993]
Ag_8SnSe_6	350,3±1,8	320,4±8,1	736,6±23,8	[Babanly, M.B., 1993]
AgTlS	72,3±0,6	72,9±3,0	136,7±8,1	[Ibragimova G.I., 2001]
	71,7±0,7	*73,3±3,2*	*133,3±5,9*	[Babanly, M.B., 1993]
$Ag_7Tl_3S_5$	320,2±2,6	351,9±15,7	544,2±36,5	[Ibragimova G.I., 2001]
	303,5±3,0	*315,0±15,0*	*614,5±30,1*	[Babanly, M.B., 1993]
Ag_3TlS_2	120,0±0,9	128,4±5,6	227,7±12,8	[Ibragimova G.I., 2001]
	114,3±1,2	*118,0±6,0*	*243,8±23,1*	[Babanly, M.B., 1993]
$Ag_8Tl_2S_5$	278,0±2,4	287,5±14,6	597,0±32,3	[Ibragimova G.I., 2001]
	266,5±2,9	*269,0±15,0*	*620,9±26,0*	[Babanly, M.B., 1993]
Ag_7TlS_4	195,9±1,5	192,0±9,4	503,2±20,2	[Ibragimova G.I., 2001]
	189,2±2,8	*184,0±13,0*	*507,1±27,1*	[Babanly, M.B., 1993]
Ag_7TlSe_4	227,1±0,4	183,4±1,9	676,7±15,1	
	234,8±1,9	198,3±6,7	652,9±27,9	[Babanly, M.B., 1982]
Ag_3TlSe_2	130,8±0,2	113,5±1,1	333,9±12,8	
	133,3±0,9	116,8±3,2	292,2±14,1	[Babanly, M.B., 1982]
AgTlSe	80,4±0,2	74,3±1,2	173,9±5,9	
	82,4±0,5	75,8±1,8	176,9±7,7	[Babanly, M.B., 1982]
$AgTlTe_2$	69,7±0,7	71,5±1,8	199,9±5,7	
	69,9±0,7	62,1±2,3	232,1±7,2	[Babanly, M.B., 1982]
$Ag_8Tl_2Te_5$	273,4±2,0	253,3±6,8	784,1±21,6	
	267,9±3,3	234,4±6,9	829,7±19,7	[Babanly, M.B., 1982]
AgTlTe	67,3±0,5	63,6±1,2	168,8±5,9	
	69,3±0,6	62,8±2,0	178,7±6,0	[Babanly, M.B., 1982]
$AgTl_3Te_2$	151,5±1,2	149,1±1,9	342,1±7,2	
	153,5±0,9	147,3±3,1	355,3±9,5	[Babanly, M.B., 1982]
Ag_9TlTe_5	238,2±1,4	206,9±6,5	800,3±19,2	
	235,6±3,4	201,5±5,9	809,5±16,6	[Babanly, M.B., 1982]
$AgAs_3Se_5$	111,9±10,5	112,5±0,5	359,1±7,5	[Babanly, M.B., 2009]
$AgAsSe_2$	55,6±3,6	54,4±3,5	167,0±3,8	[Babanly, M.B., 2009]
Ag_7AsSe_6	214±4,2	199,5±6,5	637±16	[Babanly, M.B., 2009]
Ag_3AsSe_3	107,8±3,8	101,8±4,6	310,7±7,7	[Babanly, M.B., 2009]
$AgSbS_2$	110±5	103±5	175,1±6,9	[Babanly, M.B., 1993]
Ag_3SbS_3	153±5	141±7	309,2±13,3	[Babanly, M.B., 1993]
Ag_7SbS_6	233±7	218±12	595,3±28,1	[Babanly, M.B., 1993]
$AgSbSe_2$	92,5±4,5	91,0±5	177,2±6	[Babanly, M.B., 1993]
$AgSbTe_2$	49,6±1,5	44,5±1,3	204,6±5	[Babanly, M.B., 1993]
$AgBi_3S_5$	326,4±12,9	323,4±7,7	367,5±16,5	[Shykhyev Y.M. 1995]
$AgBiS_2$	124,2±4,4	118,6±3,0	166,3±6,8	[Shykhyev Y.M. 1995]
$AgBiSe_2$	100,5±0,8	94,1±2,3	205±10	[Shykhyev Y.M. 2003]

Table 5. Standard thermodynamic functions of formation and standard entropy of some ternary chalcogenides of silver.

5. Conclusion

The results of this chapter show that Cu^+ and Ag^+ conducting superionic conductors can be successfully applied as solid-state electrolyte in the thermodynamic studies and specification of solid-phase equilibria diagrams of ternary copper - and silver containing systems by EMF method. Unlike from classical variant of EMF method with liquid electrolyte they allow to investigate also the systems containing high electrochemical active component than copper or silver (in our case, thallium). It is due to that the solid-state electrolyte, unlike liquid, prevents percolation of collateral processes (interaction of electrodes with electrolyte and through electrolyte among themselves) and by that allows to obtain reproduced data for irreversible in classical understanding of concentration chains. The specified advantage solid-state cation-containing systems by EMF methe of formation and standard entropy of some ternary chalcogenides of silverrnary phases have been conducting electrolytes allows to essentially expanding a circle of the systems investigated by EMF method.

6. Acknowledgments

The authors would like to acknowledge to Dr. Samira Z. Imamaliyeva, Dr. Ziya S. Aliev and Leyla F. Mashadiyeva and for carrying out experiments and preparation of this chapter.

7. References

Abbasov, A.S. (1981). *Thermodynamic Properties of Some Semiconductor Substances*, Elm, Baku

Babanly, M.B., Kuliyev A.A. (1982a). Phase Equilibria and Thermodynamic Properties of the Ag-Tl-Se System. *Russian Journal of Inorganic Chemistry*,Vol.27, No.9 (September), pp.1336-1340, ISSN 0036-0236

Babanly, M.B., Kuliev A.A. (1982b). Phase Equilibria and Thermodynamic Properties of the Ag-Tl-Te System, *Russian Journal of Inorganic Chemistry*, Vol. 27, No.6 (June), pp.867–872.

Babanly M.B., Kuliyev A.A. (1985a), Thermodynamic investigation and refined the phase diagram o ternary chalcogenides systems by method EMF, In: *Mathematical Problems in Chemical Thermodynamics*, Novosibirsk, Nauka, Siberian Section, pp.192-201

Babanly M.B. (1985b), Application of partial heterogeneous molar functions method in thermodynamic analysis of the ternary condensed systems, In: *Application of Mathematical Methods to studying Physico-Chemical Equilibriua*, Vol.2, Novosibirsk p.202-204

Babanly, M.B., Li Tai Un. & Kuliev, A.A. The Cu–Tl–S System (1986). *Russian Journal of Inorganic chemistry*, Vol. 32, No.7 (July), p.1837-1844, ISSN 0036-0236

Babanly, M.B., Yusibov, Yu.A. & Abishov, V.T. (1992). *Method of Electromotive Forces in the Thermodynamics of Solid Semiconductor Substances*, Baku State University.

Babanly, M.B., Yusibov, Yu.A. & Abishev, V.T. (1993). *Three-Component Chalcogenides on the Basis of Copper and Silver*, Baku, Baku State University.

Babanly, M.B., Babanly, N.B., & Mashadiyeva, L.F. (2001). *Phase Diagrams and Thermodynamic Properties of the Cu–B^{IV}(B^{V})–Chalcogen Systems*. Proceedings of VI International School-Conference "Phase Diagrams in Material Science», pp.5-6, Kiev, October, 14-20, 2001

Babanly, M.B., Shykhyev, Yu.M. &Yusibov Yu.A. (2003). Thermodynamic investigation of ternary compounds of Ag-Tl-Te system by EMF method with solid-state electrolyte. *News of Baku University, natural science series,* No.1, pp. 30-35, ISSN 1609-0586

Babanly, M.B., Mashadiyeva, L.F., & Imamaliyeva, S.Z. (2009) Thermodynamic Study of the Ag-As-Se and Ag-S-I Systems Using the EMF Method with a Solid Ag_4RbI_5 Electrolyte. *Russian Journal of Electrochemistry,* Vol.45, No.4, (April 2009), pp.399–404, ISSN 1023-1935,

Babanly, M.B., Salimov, Z.E., Babanly, N.B. & Imamalieva, S. Z. (2011). Thermodynamic properties of copper thallium tellurides. *Inorganic Materials,* Vol. 47, No.4 (April), pp.361-364, ISSN 0020-1685

Babanly, N.B., Mokhtasebzadeh, Z; Aliyev, I.I. & Babanly, M.B. (2007) Phase equilibriums in the system Cu-Bi-Te and thermodynamic properties of $CuBiTe_2$ compound. *Reports of National Academi of Science of Azerbaijan,* Vol.63, No.1, pp.41-54,

Babanly, N.B., Yusibov, Yu.A., Mirzoeva, R.J., Shykhyev, Yu.M. & Babanly, M.B. (2009). $Cu_4RbCl_3I_2$ Solid Superionic Conductor in Thermodynamic Study of Three-Component Copper Chalcogenides. *Russian Journal of Electrochemistry,* Vol. 45, No.4 (April), pp. 405–410, ISSN 1023-1935

Babanly, N.B., Aliev, Z.S., Yusibov, Yu.A.& Babanly, M.B. (2010). A Thermodynamic Study of Cu–Tl–S System by EMF Method with $Cu_4RbCl_3I_2$ Solid Electrolyte. *Russian Journal of Electrochemistry,* Vol.46, No.3 (March), pp.371-375, ISSN 1023-1935

Babanly, N.B., Yusibov, Yu.A., Aliev, Z.S. & Babanly, M.B. (2010). Phase Equilibria in the Cu–Bi–Se System and Thermodynamic Properties of Copper Selenobismuthites. *Russian Journal of Inorganic Chemistry,* Vol.55, No.9 (September), pp.1471–1481, ISSN 0036-0236

Gordon, A; Ford, R. (1972). *The chemists Companion. A handbook of practical data techniques and references.* A Wiley-Interscience Publication. New-York-London-Sydney-Toronto

Gurevich, Yu.Ya; Kharkats, Yu.I. (1992). *Superionic conductors,* Nauka, ISBN 5-02-014622-6, Moscow

Hagenmuller, P., Gool, W.V. (1978) *Solid Electrolytes, General Principles, Characterization, Materials, Applications,* Acad. Press, ISBN 0-12-313360-2, New York, San Fransisco, London

Ibragimova, G.I., Shikhiyev, Yu.M., & Babanly M.B. (2006). Solid Phase Equlibria in Ag-Ga-S (Se,Te) systems and thermodynamic properties of ternary phases. *Chemical Problems*, No.1, pp.23-28, ISBN 9952-8034-4-6

Ibragimova, G.I., Babanly, M.B., & Shykhyev Yu.M. (2001). *Investigation of system Ag_2Te-Ga_2Te_3 by EMF method.* Proceedings of VII National Science conference "Physico-chemical analyses and inorganic material science", pp.41-45, Baku State University, Baku, May 21-23, 2001

Ibragimova, G.I., Shykhyev Yu.M. & Babanly, M.B. *Thermodynamic investigation of system Ag_2S-Tl_2S-S.* Proceedings of VII National Science conference "Physico-chemical analyses and inorganic material science", pp.3-10, Baku State University, Baku, May 21-23, 2001/

Ivanov-Shits, A.K. & Murin, I.V. (2000). *Solid–State Ionics*, Sankt-Peterburg State University, ISBN 5-288-02745-5, Sankt-Petersburg

Kornilov, A.N. Stepina L.B., Sokolov V.A. (1972). The recommendation for the compact form of presents of experimental data at the publication of results of thermochemical and thermodynamic invstigations. *Russian Journal of Physical Chemistry,* Vol.46, No.11 (November), pp.2974-2980, ISSN 0036-0244

Kubaschewski, O., Alcock, C.B., Spenser P.J. (1993). *Materials Thermochemistry*. Pergamon Press, INSB 0080418899

Mills, K. (1974). *Termodinamic data for inorganic sulphids, selenides and tellurides.* Butterworths, ISBN 10 040870537X, London.

Morachevskii, A.G., Voronin, G.F., & Kutsenok, I.B. (2003). *Electrochemical Research Methods in Thermodynamics of Metallic Systems,* ITsK "Akademkniga", ISBN 5-94628-064-3, Moscow:

Shevelkov, A.V. (2008). Chemical aspects of the design of thermoelectric materials. *Russian Chemical Reviews,* Vol. 77, No.1 (January), pp.1-19, ISSN: 0036-021X

Shykhyev, Yu.M., Yusibov, Yu.A., Popovkin, B.A., Babanly, M.B. (2003).The Ag-Bi-Se System. *Russian Journal of Inorganic Chemistry,* Vol. 48, No.12 (December), pp.1941-1948, ISSN 0036-0236

Shykhyev, Yu.M., Yusibov, Yu.A., Babanly, M.B. (1995). Thermodynamic properties of system Ag-Bi-S. *News of Baku University, natural science series,* No.1, pp.93-98, ISSN 1609-0586

Ternary Alloys. A Comprehensive Compendium of Evaluated Constitutional Data and Phase Diagrams. V.1-5, Max Plank In-t, Stuttgart, 1992-1995

Yungman, V.S. (2006) Database of Thermal Constants of Compounds, Electronic Version, httr://www.chem.msu.su/sgi-bin/tkv.

Vasil'yev, V.P., Nikol'skaya, A.V., Gerasimov, Ya.,I., Kuznetsov A.F. (1968). Thermodynamic investigation of thallium tellurides by method EMF. *Inorganic Materials,* Vol.4, No.7, pp. 1040-1046, ISSN 0020-1685.

Voronin, G.A. (1976) Partial thermodynamic functions of heterogeneous alloys and their application in thermodynamics of alloys. In: *Modern problems of physical chemistry,* pp.29-48. Moscow State University, Moscow.

Wagner, C. (1952). *Thermodynamics of Alloys*. Addison-Wesley Press, Cambridge.

West, A.R. (1987). *Solid State Chemistry and Its Applications.* Wiley, ISBN 978-0-471-90874-6, Camden, NY, U.S.A.

Part 2

Application of Electromotive Force

5

Electromotive Force Measurements and Thermodynamic Modelling of Sodium Chloride in Aqueous-Alcohol Solvents

I. Uspenskaya, N. Konstantinova, E. Veryaeva and M. Mamontov
Lomonosov Moscow State University
Russia

1. Introduction

In chemical engineering, the liquid extraction plays an important role as a separation process. In the conventional solvent extraction, the addition of salts generally increases the distribution coefficients of the solute and the selectivity of the solvent for the solute. Processes with mixed solvent electrolyte systems include regeneration of solvents, extractive crystallization, and liquid-liquid extraction for mixtures containing salts. For instance, combining extraction and crystallization allowed effective energy-saving methods to be created for the isolation of salts from mother liquors (Taboada et al., 2004), and combining extraction with salting out and distillation led to a new method for separating water from isopropanol (Zhigang et al., 2001). Every year a great financial support is required for conceptual design, process engineering and construction of chemical plants (Chen, 2002). Chemical engineers perform process modeling for the cost optimization. Success in that procedure is critically dependent upon accurate descriptions of the thermodynamic properties and phase equilibria of the concerned chemical systems.
So there is a great need in systematic experimental studies and reliable models for correlation and prediction of thermodynamic properties of aqueous-organic electrolyte solutions. Several thermodynamic models have been developed to represent the vapor-liquid equilibria in mixed solvent-electrolyte systems. Only a few studies have been carried out concerning solid-liquid, liquid-liquid and solid-liquid-vapor equilibrium calculations. The lists of relevant publications are given in the reviews of Liddell (Liddell, 2005) and Thomsen (Thomsen et al., 2004); some problems with the description of phase equilibria in systems with strong intermolecular interactions are discussed in the same issues. Among the problems are poor results for the simultaneous correlation of solid - liquid - vapor equilibrium data with a single model for the liquid phase. This failure may be due to the lack of reliable experimental data on thermodynamic properties of solutions in wide ranges of temperatures and compositions. Model parameters were determined only from the data on the phase equilibrium conditions in attempt to solve the inverse thermodynamic problem, which, as is known, may be ill-posed and does not have a unique solution (Voronin, 1992). Hence, the introduction of all types of experimental data is required to obtain a credible thermodynamic model for the estimation of both the thermodynamic functions and equilibrium conditions. One of the most reliable methods for the

determination of the activity coefficients of salts in solutions is the Method of Electromotive Force (EMF).

The goal of this work is to review the results of our investigations and literature data about EMF measurements with ion-selective electrodes for the determination of the partial properties of some salts in water-alcohol mixtures. This work is part of the systematic thermodynamic studies of aqueous-organic solutions of alkali and alkaline-earth metal salts at the Laboratory of Chemical Thermodynamics of the Moscow State University (Mamontov et al., 2010; Veryaeva et al., 2010; Konstantinova et al., 2011).

2. Ion-selective electrodes in the thermodynamic investigations

The measurement of the thermodynamic properties of aqueous electrolyte solutions is a part of the development of thermodynamic models and process simulation. There are three main groups of experimental methods to determine thermodynamic properties, i.e., calorimetry, vapour pressure measurements, and EMF measurements. The choice of the method is determined by the specific properties of the studied objects, and purposes which are put for the researcher. EMF method and its application for thermodynamic studies of metallic and ceramic systems has been recently discussed in detail by Ipser et al. (Ipser et al., 2010). The use of this technique in the thermodynamics of electrolyte solutions is described in many books and articles. In this paper we focus on the determination of partial and integral functions of electrolyte solutions using electrochemical cells with ion-selective electrodes (ISE).

Some background information on ISE may be found in Wikipedia. According to the definition given there an ion-selective electrode is a transducer (or sensor) that converts the activity of a specific ion dissolved in a solution into an electrical potential, which can be measured by a voltmeter or pH meter. The voltage is theoretically dependent on the logarithm of the ionic activity, according to the Nernst equation. The main advantages of ISE are good selectivity, a short time of experiment, relatively low cost and variety of electrodes which can be produced.

The principles of ion-selective electrodes operation are quite well investigated and understood. They are detailed in many books; for instance, see the excellent review of Wroblewski (http://csrg.ch.pw.edu.pl/tutorials/ise/). The key component of all potentiometric ion sensors is an ion-selective membrane. In classical ISEs the arrangement is symmetrical which means that the membrane separates two solutions, the test solution and the inner solution with constant concentration of ionic species. The electrical contact to an ISE is provided through a reference electrode (usually Ag/AgCl) implemented in the internal solution that contains chloride ions at constant concentration. If only ions penetrate through a boundary between two phases - a selective membrane, then as soon as the electrochemical equilibrium will be reached, the stable electrical potential jump will be formed. As the equilibrium potential difference is measured between two identical electrodes placed in the two phases we say about electromotive force. Equilibrium means that the current of charge particles from the membrane into solution is equal to the current from the solution to the membrane, i.e. a potential is measured at zero total current. This condition is only realized with the potentiometer of high input impedance (more than 10^{10} Ohm). In the case of the ion selective electrode, EMF is measured between ISE and a reference electrode, placed in the sample solution. If the activity of the ion in the reference phase (a_{ref}) is kept constant, the unknown activity of component in solution under investigation (a_X) is related to EMF by Nernst equation :

$$E = E_0 + \frac{RT}{nF} \ln \frac{a_X}{a_{\mathrm{ref}}} = const + S \log a_X \,, \tag{1}$$

where E_0 is a standard potential, S is co-called, Nernst slope, which is equal to $59.16/n$ (mV) at 298.15 K and n - the number of electrons in Red/Ox reaction or charge number of the ion X (z_X). Ions, present in the sample, for which the membrane is impermeable, will have no effect on the measured potential difference. However, a membrane truly selective for a single type of ions and completely non-selective for other ions does not exist. For this reason the potential of such a membrane is governed mainly by the activity of the primary ion and also by the activity of other ions. The effect of interfering species Y in a sample solution on the measured potential difference is taken into consideration in the Nikolski-Eisenman equation:

$$+S \log(a_x + K_{xy} a_y^{\frac{z_x}{z_y}}) \,, \tag{2}$$

where a_Y is the activity of ion Y, z_Y its charge number and K_{xy} the selectivity coefficient (determined empirically). The values of these coefficients for ISE are summarized, for example, in the IUPAC Technical Reports (Umezawa et al., 2000, 2002).

The properties of an ion-selective electrode are characterized by parameters like selectivity, slope of the linear part of the measured calibration curve of the electrode, range of linear response, response time and the temperature range. Selectivity is the ability of an ISE to distinguish between the different ions in the same solution. This parameter is one of the most important characteristics of an electrode; the selectivity coefficient K_{XY} is a quantative measure of it. The smaller the selectivity coefficient, the less is the interference of the corresponding ion. Some ISEs cannot be used in the presence of certain other interfering ions or can only tolerate very low contributions from these ions. An electrode is said to have a Nernstian response over a given concentration range if a plot of the potential difference (when measured against a reference electrode) versus the logarithm of the ionic activity of a given species in the test solution, is linear with a slope factor which is given by the Nernst equation, i.e. $2.303RT/nF$. The slope gets lower as the electrode gets old or contaminated, and the lower the slope the higher the errors on the sample measurements. Linear range of response is that range of concentration (or activity) over which the measured potential difference does not deviate from that predicted by the slope of the electrode by more that ± 2 mV. At high and very low ion activities there are deviations from linearity; the range of linear response is presented in ISE passport (typically, from 10^{-5} M to 10^{-1} M). Response time is the length of time necessary to obtain a stable electrode potential when the electrode is removed from one solution and placed in another of different concentration. For ISE specifications it is defined as the time to complete 90% of the change to the new value and is generally quoted as less than ten seconds. In practice, however, it is often necessary to wait several minutes to complete the last 10% of the stabilization in order to obtain the most precise results. The maximum temperature at which an ISE will work reliably is generally quoted as 50°C for a polymeric (PVC) membrane and 80°C for crystal or 100°C for glass membranes. The minimum temperature is near 0°C.

The three main problems with ISE measurements are the effect of interference from other ions in solution, the limited range of concentrations, and potential drift during a sequence of measurements. As known, the apparent selectivity coefficient is not constant and depends on several factors including the concentration of both elements, the total ionic strength of the solution, and the temperature. To obtain the reliable thermodynamic information from the results of EMF measurements it's necessary to choose certain condition of an experiment to avoid the interference of other ions. The existence of potential drift can be observed if a series of standard solutions are repeatedly measured over a period of time. The results show that the difference between the voltages measured in the different solutions remains essentially the same but the actual value generally drifts in the same direction by several millivolts. One way to improve the reliability of the EMF measurements is to use multiple independent electrodes for the investigation the same solution.

Due to the limited size of this manuscript we cannot describe in detail the history of ISE and their applications in physical chemistry. For those interested, we recommend to read the reviews (Pungor, 1998; Buck & Lindner, 2001; Pretsch, 2002; Bratov et al., 2010). The application of ion-selective electrodes in nonaqueous and mixed solvents to thermodynamic studies was reviewed by Pungor et al (Pungor et al., 1983), Ganjali and co-workers (Ganjali et al., 2007) and Nakamura (Nakamura, 2009). In the end of the XX-th century the results of systematic thermodynamic investigations with ISEs were intensively published by Russian (St. Petersburg State University and Institute of Solution Chemistry, Russian Academy of Sciences) and Polish scientists from the Lodz University. At the present time the systematic and abundant publications in this branch of science belong to the Iranian investigators (Deyhimi et al., 2009; 2010). The latter group are specialized in the development of many sensors, and particularly, carrier-based solvent polymeric membrane electrodes for the determination of activity coefficients in mixed solvent electrolyte solutions. Studies of the thermodynamic properties of salts in mixed electrolytes by EMF are also being conducted by Portuguese, Chinese and Chilean scientists.

3. Thermodynamic models for mixed solvent–electrolyte systems

It is well known that nonideality in a mixed solvent–electrolyte system can be handled using expression for the excess Gibbs energy (G^{ex}, J). According to Lu and Maurer (Lu & Maurer, 1993), thermodynamic models for aqueous electrolytes and electrolytes in mixed solvents are classified as either physical or chemical models. The former are typically based on extensions of the Debye–Hückel equation, the local composition concept, or statistical thermodynamics. As some examples of first group of models should be mentioned the Pitzer model (Pitzer & Mayorga, 1973) and its modifications - Pitzer-Simonson (Pitzer & Simonson, 1986) and Pitzer-Simonson-Clegg (Clegg & Pitzer, 1992); eNRTL (Chen et al., 1982; Mock et al., 1986) and its variants developed by scientists at Delft University of Technology (van Bochove et al., 2000) or Wu with co-workers (Wu et al., 1996) and Chen et al (Chen et al., 2001; Chen & Song, 2004); eUNIQUAC (Sander et al., 1986; Macedo et al., 1990; Kikic et al. 1991; Achard et al. , 1994). The alternative chemically based models assume that ions undergo solvation reactions. The most important examples of that group of thermodynamic models are model of Chen (Chen et al, 1999) and chemical models have been proposed by Lu and Maurer (1993), Zerres and Prausnitz (1994), Wang (Wang et al., 2002; 2006) and recently developed COSMO-SAC quantum mechanical model, a variation of COSMO-RS (Klamt, 2000; Lin & Sandler, 2002). The detailed description of those models is given in original papers, the brief reviews are presented by Liddell (Liddell, 2005) and Smirnova (Smirnova, 2003). According to Chen (Chen, 2006), the perspective models of electrolytes in mixed solvents would require no ternary parameters, be formulated in the

concentration scale of mole fractions, represent a higher level of molecular insights, and preferably be compatible with existing well-established activity coefficient models.
The excess Gibbs free energy per mole of real solution comprises three (sometimes, four) terms

$$G^{ex} = G^{ex,\,lr} + G^{ex,\,sr} + G^{ex,\,Born}. \tag{3}$$

The first term represents long range (lr) electrostatic forces between charged species. Usually the unsymmetric Pitzer-Debye-Hückel (PDH) model is used to describe these forces. The second contribution represents the short-range (sr) Van der Waals forces between all species involved. The polynomial or local composition models, based on reference states of pure solvents and hypothetical, homogeneously mixed, completely dissociated liquid electrolytes are applied to represent a short-range interactions. The model is then normalized by infinite dilution activity coefficients in order to obtain an unsymmetric model. And the third term is a so-called the Born or modified Brönsted–Guggenheim contribution. The Born term is used to account for the Gibbs energy of transfer of ionic species from an infinite dilution state in a mixed solvent to an infinite dilution state in the aqueous phase; for the electrolyte MX of 1,1-type:

$$\frac{G^{ex,Born}}{RT} = \frac{x_{MX} e^2}{8\pi\varepsilon_0 k_B T}\left(\frac{1}{\varepsilon_s} - \frac{1}{\varepsilon_w}\right)\left(\frac{1}{r_M} + \frac{1}{r_X}\right) \tag{4}$$

where ε_s and ε_w are the relative dielectric constants of the mixed solvent and water, respectively, ε_0 is the electric constant, k_B is the Boltzmann constant and r_M, r_X are the Born radii of the ions (Rashin&Honig, 1985), e is the electron charge. With the addition of the Born term, the reference for each ionic species will always be the state of infinite dilution in water, disregarding the composition of the mixed solvent.
The most frequently and successful model used to describe the thermodynamic properties of aqueous electrolyte solutions is the ion interaction or virial coefficient approach developed by Pitzer and co-workers (Pitzer & Mayorga, 1973). In terms of Pitzer formalizm, the mean ionic activity coefficient of the 1,1-electrolyte in the molality scale ($\gamma_\pm$) is determined according to the following equation:

$$\ln(\gamma_\pm) = -A_\phi\left[\frac{m^{1/2}}{1+1.2m^{1/2}} + \frac{2}{1.2}\ln\left(1+1.2m^{1/2}\right)\right] + B^{\gamma}_{MX} m + C^{\gamma}_{MX} m^2, \tag{5}$$

$$A\phi = \frac{\left(2\pi\rho_s N_A\right)^{1/2}}{3}\left(\frac{e^2}{4\pi\varepsilon_0 \varepsilon k_B T}\right)^{3/2}, \tag{5a}$$

$$B^{\gamma}_{MX} = 2\beta^{(0)}_{MX} + \frac{\beta^{(1)}_{MX}}{2m}\left[1-\exp\left(-2m^{1/2}\right)\left(1+2m^{1/2}-2m\right)\right],\ C^{\gamma}_{MX} = const, \tag{5b}$$

The osmotic coefficient (ϕ) of the solvent, the excess Gibbs energy (G^{ex}), and the relative (excess) enthalpy of the solution (L), can be calculated as:

$$1-\phi = A_\phi\left[\frac{m^{1/2}}{1+1.2m^{1/2}}\right] - m\left[\beta^{(0)}_{\mathrm{MX}} + \beta^{(1)}_{\mathrm{MX}}\exp\left(-2m^{1/2}\right)\right] - \frac{2}{3}C^{\gamma}_{\mathrm{MX}}m^2, \tag{6}$$

$$G^{ex} = 2n_s mRT\left(1-\phi+\ln(\gamma_\pm)\right), \tag{7}$$

$$L = -T^2\left(\partial\left(G^{ex}/T\right)/\partial T\right)_{p,m} = 2mRT^2\left(\left(\partial\phi/\partial T\right)_{p,m} - \left(\partial\ln\left(\gamma_\pm\right)/\partial T\right)_{p,m}\right) = \\ = 2mRT^2\left[\begin{array}{l}\left(\dfrac{\partial A_\phi}{\partial T}\right)_{p,m}\left(\dfrac{2}{1.2}\ln\left(1+1.2m^{1/2}\right)\right) - m\left(\dfrac{\partial\beta^{(0)}_{\mathrm{MX}}}{\partial T}\right)_{p,m} + \\ +\dfrac{1}{2}\left(\dfrac{\partial\beta^{(1)}_{\mathrm{MX}}}{\partial T}\right)_{p,m}\left(\exp\left(-2m^{1/2}\right)\left(1+2m^{1/2}\right)-1\right) + \dfrac{m^2}{2}\left(\dfrac{\partial C^{\gamma}_{\mathrm{MX}}}{\partial T}\right)_{p,m}\end{array}\right]. \tag{8}$$

In the above equations, A_ϕ is the Debye–Hückel coefficient for the osmotic function, ρ_s is the density of solvent, N_A is the Avogadro's number, n_s is the weight of the solvent (kg), M_s is the molar mass of the solvent (g·mol^{-1}), $\beta^{(0)}_{\mathrm{MX}}$, $\beta^{(1)}_{\mathrm{MX}}$, and C^{γ}_{MX} are model parameters characterizing the binary and ternary interactions between ions in the solution. The densities and dielectric constants of the mixed solvent can be obtained experimentally or calculated in the first approximation as

$$\rho_{\mathrm{s}} = \rho_n\frac{M_{\mathrm{s}}}{\left(\sum_n x'_n M_n\right)}, \quad \varepsilon_{\mathrm{s}} = \sum_n \varphi_n\varepsilon_n \tag{9}$$

where x'_n is the salt-free mole fraction of solvent n in the solution, V_n is the molar volume of the pure solvent n, φ_n is the volume fraction of solvent n. Molar mass and the volume fraction of the mixed solvent are represented as

$$M_{\mathrm{s}} = \sum_n x'_n M_n, \quad \varphi_n = \frac{x'_n V_n}{\sum_m x'_m V_m} \tag{10}$$

The Pitzer model is widely used but it has some drawbacks: (a) it requires both binary and ternary parameters for two-body and three-body ion–ion interactions; (b) the Pitzer model has the empirical nature, as a result there are some problems with the description of temperature dependences of binary and ternary parameters; (c) it is formulated in the basis of the concentration scale of molality. In practice, the Pitzer model and other similar molality-scale models can only be used for dilute and middle concentration range of aqueous electrolyte systems. Pitzer, Simonson and Clegg (Pitzer & Simonson, 1986; Clegg & Pitzer, 1992) proposed a new version of the Pitzer model, developed at the mole fraction base that can be applied to concentrations up to the pure fused salt for which the molality is infinite. The short-range force term is written as the Margules expansion:

$$\frac{G^{\mathrm{ex,\,sr}}}{RT} = \sum_j\sum_i x_i x_j\left[w_{ij} + u_{ij}(x_i - x_j)\right] + \sum_k\sum_j\sum_i x_i x_j x_k C_{ijk} - \frac{G^0_S}{RT}, \tag{11}$$

where w_{ij}, u_{ij}, and C_{ijk} are binary and ternary interaction coefficients, respectively, x_i is the mole fraction of the species i in the mixture. The last term in Eq.(11) is introduced to account various possible types of reference states for electrolyte in solution - pure fused salt and state of infinite dilution. The contribution from the long-range forces, i.e. Debye–Hückel interactions, is given by

$$\frac{G_{DH}^{ex}}{RT} = -\frac{4A_x I_x}{\rho}\ln\left(\frac{1+\rho I_x^{1/2}}{1+\rho\left(I_x^0\right)^{1/2}}\right) + \sum_M\sum_X x_M x_X B_{MX} g(\alpha I_x^{1/2}), \quad (12)$$

$$\text{with } g(y) = 2\left[1-(1+y)e^{-y}\right]/y^2 \quad (12a)$$

In the above equations x_M, x_X are molar fractions of ionic species in solution; A_x is the Debye–Hückel coefficient for the osmotic function at the molar fraction basis ($A_x = A_\phi / M_s^{1/2}$); I_x is the ionic strength on the mole fraction basis which, for single-charged ions, $I_x = 0.5(x_M + x_X)$; I_x^0 is the ionic strength of solution in a standard state of the pure fused salt, it approaches 0 at infinite dilution for the asymmetric reference state. The parameter ρ is equivalent to the distance of closest approach in the Debye–Hückel theory, both parameters, ρ and α, in Eq.(12) are equal to 13 for 1,1-electrolyties. B_{MX} is a specific parameter for each electrolyte. For a mixture of two neutral species, 1 and 2, and a strong 1:1 electrolyte MX wth the reference state of infinite dilution, the contributions of the short and long-range force terms to the mean ionic activity coefficient of the electrolyte MX, at the mole fraction basis can be written as follows:

$$\ln\gamma_x = \ln\gamma_x^{sr} + \ln\gamma_x^{DH}, \quad (13)$$

$$\begin{aligned}\ln\gamma_x^{DH} = &-A_x\left[\frac{2}{\rho}\ln(1+\rho\sqrt{I_x}) + \frac{(1-2I_x)\sqrt{I_x}}{1+\rho\sqrt{I_x}}\right] + x_X B_{MX} g(\alpha\sqrt{I_x}) - \\ &- x_M x_X B_{MX}\left[\frac{g(\alpha\sqrt{I_x})}{2I_x} + \left(1-\frac{1}{2I_x}\right)e^{-\alpha\sqrt{I_x}}\right].\end{aligned}, \quad (13a)$$

$$\begin{aligned}\ln\gamma_x^{sr} = &\frac{x_1 x_2}{f^2}\left\{\left(1-f^2\right)w_{12} + 2(x_1-x_2)\frac{1-f^3}{f}u_{12} + \left[\left(1-2x_I\right)f^2 - 1\right]Z_{12MX}\right\} + \\ &+ \frac{f^2-1}{f}\left(x_1 W_{1MX} + +x_2 W_{2MX}\right) + \frac{x_1}{3f^2}\left[f^3\left(2-2x_1+x_I\right) + x_I f^2\left(3x_1+x_2\right) - 2x_2\right]U_{1MX} + \\ &+ \frac{x_2}{3f^2}\left[f^3\left(2-2x_2+x_I\right) + +x_I f^2\left(3x_2+x_1\right) - 2x_1\right]U_{2MX}\end{aligned} \quad .(13b)$$

In the above equations $x_I = 2x_M = 2x_X = 1 - x_1 - x_2$, $f = 1 - x_I$; w_{12} and u_{12} are model parameters for the binary system (solvent 1 and solvent 2), W_{iMX} and U_{iMX} are model parameters for the binary system - solvent i with MX (i = 1 or 2), Z_{12MX} is a model parameter which accounts for the triple interaction. Formula details can be found in (Lopes et al, 2001).

The approximation of the experimental data in the present study was carried out with the Pitzer and Pitzer-Simonson models. The result of this investigation can be used in future for the development of a new thermodynamic models and verification of existing ideas.

4. EMF measurements of galvanic cells with ternary solutions NaCl – H_2O – $C_nH_{2n+1}OH$ (n = 2-5). Experimental procedure

Sodium chloride (reagent grade, 99.8%) was used in experiments. The salt was additionally purified by the double crystallization of NaCl during evaporation of the mother liquor. The purified salt was dried *in vacuo* at 530 K for 48 h. The isomers of alcohol were used as organic solvents: C_2H_5OH (reagent grade, 99.7%), *1*-C_3H_7OH (special purity grade, 99.94 %), *iso*-C_3H_7OH (reagent grade, 99.2%), *1*-C_4H_9OH (special purity grade, 99.99%) and *iso*-C_4H_9OH (reagent grade, 99.5%), *1*-$C_5H_{11}OH$ (reagent grade, 99.6%) and *iso*-$C_5H_{11}OH$ (reagent grade, 99.5%). To remove moisture, the alcohols were kept on zeolites 4A for 7 days and then distilled under the atmospheric pressure. The purity of alcohols was confirmed by the agreement of the measured boiling points of the pure solvents at atmospheric pressure and the refractive indices with the corresponding published data. Deionized water with a specific conductance of 0.2 μS cm^{-1} used in experiments was prepared with a Millipore Elix filter system.

Electrochemical measurements were carried out with the use of solutions of sodium chloride in mixed water-organic solvents at a constant water-to-alcohol weight ratio. The reagents were weighed on a Sartorius analytical balance with an accuracy of ±0.0005 g. A sample of the NaCl was transferred to a glass cell containing ~30 g of a water-alcohol solution. The cell was tightly closed with a porous plastic cap to prevent evaporation of the solution. The cell was temperature-controlled in a double-walled glass jacket, in which the temperature was maintained by circulating water. The temperature of the samples was maintained constant with an accuracy of ±0.05 K. The solutions were magnetically stirred for 30 min immediately before the experiments. All electrochemical measurements were carried out in the cell without a liquid junction; the scheme is given below (I):

$$Na^+\text{-ISE} \mid NaCl(m)+H_2O(100-w_{alc})+(1\text{- or } iso\text{-})C_nH_{2n+1}OH(w_{alc}) \mid Cl^-\text{-ISE}, \qquad \text{(I)}$$

where w_{alc} is the weight fraction of alcohol in a mixed solvent expressed in percentage and m is the molality of the salt in the ternary solution. Concentration cell (I) EMF measurements are related to the mean ionic activity coefficient by the equation

$$E = E_0 - \frac{2RT}{F}\ln(\gamma_{\pm}m). \qquad (14)$$

An Elis-131 ion-selective electrode for chloride ions (Cl^--ISE) was used as the reference electrode. The working concentration range of the Cl^--ISE electrode at 293 K is from $3{\cdot}10^{-5}$ to 0.1 mol·L^{-1}; the pH of solutions should be in the range from 2 to 11. A glass ion-selective electrode for sodium ions (Na^+-ISE) served as the indicator electrode, which reversibly responds to changes in the composition of the samples under study. The ESL-51-07SR (Belarus) and DX223-Na^+ (Mettler Toledo) ion-selective electrodes were used in experiments. Different concentration ranges of solutions were detected by two glass electrodes. The concentration range for the ESL-51-07SR electrode at 293.15 K is from 10^{-4} to 3.2 mol·L^{-1}; the same for the Mettler electrode is lower (from 10^{-6} to 1 mol·L^{-1}). The results of experiments performed with the use of two different ion-selective electrodes for Na^+ can be considered as independent. This increases the statistical significance of the EMF values and provides information on their correctness. The potential of the cell was measured with the use of the Multitest IPL-103 ionomer. The input impedance of the ionomer was at least 10^{12} Ohm.

The concentration range of the working solution was determined by two factors. According to the manufacturer, the sodium electrode was more sensitive to protons than to sodium ions and the interfering effect of hydrogen ions can be ignored if $[Na^+]/[H^+] > 3000$ in the solution under consideration (*i.e.*, at (pH - pNa) > 3.5). Therefore, the lower limit of molality in each series of solutions had to be no less than 0.03 $mol \cdot kg^{-1}$. To meet the condition of solution homogeneity, the highest concentration of NaCl had to be no higher than the solubility of the salt in the mixed solvent. The widest ranges of concentration were investigated in the systems with ethanol and 1-(*iso-*)propanol. All studied systems belong to the class of systems with the top critical point; the area of existence of solutions narrows with temperature increasing. To maintain the homogeneity of mixtures, the concentration of the salt in solutions was kept no higher than the upper solubility limit of sodium chloride in a mixed solvent of a given composition. In each series of experiments successive measurements were carried out for samples with constant ratios of water/alcohol components and different molalities starting with the lowest concentration. The choice of composition range is illustrated at Fig.1 where the fragment of Gibbs-Roseboom triangle of H_2O-C_2H_5OH-NaCl system at 298.15 K is drawn.

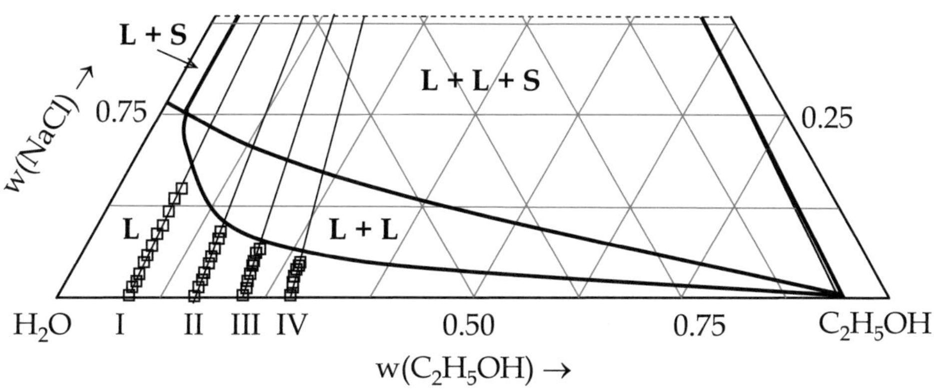

Fig. 1. Fragment of the isothermal (298.15 K) section of sodium chloride - water - ethanol system. Symbols correspond to composition of solutions under investigation. Numbers I, II, III and IV are denoted to solutions with fixed water-alcohol ratio.

Symbols are met the case of real experiment with various sodium chloride molality and constant water-to-alcohol ratios. The symbols L and S denote the Liquid and Solid phases of the ternary system. Each series of solutions was studied at two or three temperatures. The EMF values of the cell were assumed to be equilibrium if the rate of the drift in EMF was no higher than 0.01-0.02 $mV \cdot min^{-1}$. The equilibrium was established, less than 30 min. The constancy of the composition of the solution was confirmed by the fact that the refractive indices measured before and after electrochemical experiments remained practically coincided. It appeared that a change in the weight fraction of the organic solvent during experiments was at most 0.06 wt.%.

In the first step, the operation of the electrochemical cell (I) containing an aqueous sodium chloride solution (w_{alc} = 0) was tested at 288.15, 298.15, and 318.15 K. The mean ionic activity coefficients of NaCl are consistent with the published data (Silvester & Pitzer, 1977; Truesdell, 1968; Pitzer & Mayorga, 1973; Lide, 2007-2008) within an experimental error. In the second step, electrochemical measurements with the use of sodium chloride solutions in water-alcohol solvents were carried out at various temperatures. Each composition was measured at least two times using the ISE-Na^+ and ISE-Cl^- ion-selective electrodes.

5. Thermodynamic properties of solutions in the NaCl – H_2O – $C_nH_{2n+1}OH$ (n = 2-5) system

5.1 The mean ionic activity coefficient of the sodium chloride in the ternary solutions

The temperature-concentration dependence of the mean ionic activity coefficient of NaCl was approximated using the Pitzer and Pitzer-Simonson model for 1:1 electrolytes. As mentioned above, the electromotive force of the electrochemical cell (I) is related to the activity coefficient of the salt by Eq.(14). In the present work the mean ionic activity coefficient of the salt was calculated with the asymmetric normalization, in which a mixed solvent with a fixed water-to-alcohol ratio and the extremely dilute sodium chloride solution in this mixed solvent served as the standard state of the components of the solutions. The Debye–Hückel coefficient for the osmotic function (A_ϕ) was calculated by Eq. (5a) taking into account the experimental data on the densities and dielectric constants of water and the alcohols in the temperature range under consideration, which were published in (Bald et al., 1993; Frenkel et al., 1998; Balaban et al., 2002; Lide, 2007-2008; Pol & Gaba, 2008; Omrani et al., 2010). The A_ϕ values used for each composition of the solvent at different temperatures are given in Table 1.

The numerical values of the parameters E_0, $\beta^{(0)}_{\text{NaCl}}$, $\beta^{(1)}_{\text{NaCl}}$, and C^{γ}_{NaCl} were determined by the approximation of the experimental EMF values for each series of measurements. For all the solutions under consideration, the parameter C^{γ}_{NaCl} was insignificant and the error in its determination was higher than the absolute value. Hence, we assumed C^{γ}_{NaCl} = 0. It's known that the term C^{γ}_{NaCl} in Eq. (5) makes a considerable contribution only at high concentrations of the electrolyte, so this approach is valid for systems characterized by a narrow range of the existence of ternary solutions.

The calculated Pitzer model parameters and the errors in their determination (rms deviations) for the aqueous solution of sodium chloride are given in Table 2. The rms dispersions for each series of experimental data are listed in the last column. The parameters of Eq. (5) and (14) are given with an excess number of significant digits to avoid the error due to the rounding in further calculations. As can be seen from the Table 2, the model descriptions of water-salt solutions in both cases are similar, parameters consist within errors. So all experimental data for ternary mixtures were described taking into account the parameters E_0, $\beta^{(0)}_{\text{NaCl}}$, and $\beta^{(1)}_{\text{NaCl}}$ (see Table 3). Some of those results were published earlier (Mamontov et al., 2010; Veryaeva et al., 2010), thermodynamic assessment of NaCl-H_2O-*1*(or *iso*-)C_4H_9OH system is accepted to the Fluid Phase Equilibria journal.

System	w_{Alc}, wt. %	A_ϕ		
		288.15 K	298.15 K	318.15 K
NaCl - H_2O	0	0.3856	0.3917	0.4059
NaCl - H_2O - C_2H_5OH	9.99	0.4232	0.4300	0.4465
	19.98	0.4690	0.4770	0.4953
	39.96	0.5942	0.6047	0.6290
NaCl - H_2O - *1*-C_3H_7OH	9.82	-	0.4437	0.4631
	19.70	-	0.5121	0.5366
	29.62	-	0.6021	0.6339
	39.56	-	0.7231	0.7664
NaCl - H_2O - *iso*-C_3H_7OH	10.00	-	0.4466	-
	20.00	-	0.5207	-
	30.00	-	0.6191	-
	40.00	-	0.7528	-
	49.90	-	0.9373	-
	58.50	-	1.1581	-
NaCl - H_2O - *1*-C_4H_9OH	3.00	0.3958	0.4035	0.4204
	4.49	0.4024	0.4104	0.4279
	5.66	0.4079	0.4161	0.4359
NaCl - H_2O - *iso*-C_4H_9OH	3.00	0.3999	0.4119	0.4394
	4.50	0.4103	0.4231	0.4510
	5.66	0.4187	0.4315	0.4597
NaCl - H_2O - *1*-$C_5H_{11}OH$	2.00	0.3866	0.3929	-
NaCl - H_2O - *iso*-$C_5H_{11}OH$	2.00	0.3860	0.3926	-

Table 1. Debye–Hückel coefficients for the osmotic function in NaCl-H_2O-(*1-, iso-*) $C_nH_{2n+1}OH$ solutions.

T, K	$-E_0\times 10^3$, V	$\beta^{(0)}_{NaCl}$, kg·mol^{-1}	$\beta^{(1)}_{NaCl}$, kg·mol^{-1}	C^{γ}_{NaCl}, kg^2·mol^{-2}	$s_0(E)\times 10^4$
288.15	113.7 ± 0.2	0.0766 ± 0.002	0.2177 ± 0.02	0	1.2
	113.4 ± 0.2	0.0653 ± 0.007	0.2672 ± 0.03	0.0037 ± 0.002	0.8
298.15	116.1 ± 0.3	0.0838 ± 0.002	0.2285 ± 0.03	0	1.6
	115.9 ± 0.3	0.0720 ± 0.010	0.2802 ± 0.05	0.0039 ± 0.003	1.2
318.15	123.3 ± 0.5	0.0891 ± 0.004	0.2621 ± 0.04	0	2.8
	123.1 ± 0.7	0.0780 ± 0.021	0.3109 ± 0.10	0.0036 ±0.007	2.6

Table 2. Pitzer parameters for solutions of sodium chloride in water.

w_{Alc}, wt. %	m, mol·kg^{-1}	T, K	$-E_0\times 10^3$, V	$\beta^{(0)}_{NaCl}$, kg·mol^{-1}	$\beta^{(1)}_{NaCl}$, kg·mol^{-1}	$s_0(E)\times 10^4$
1	2	3	4	5	6	7
9.99 % C_2H_5OH	0.050 - 2.999	288.15	136.1 ± 0.4	0.0830 ± 0.003	0.1919 ± 0.04	2.0
		298.15	138.6 ± 0.6	0.0884 ± 0.004	0.2468 ± 0.05	2.8
		318.15	145.5 ± 0.8	0.0927 ± 0.006	0.2976 ± 0.07	4.3
19.98 % C_2H_5OH	0.050 - 2.998	288.15	158.3 ± 0.	0.0861 ± 0.004	0.1567 ± 0.04	2.5
		298.15	160.7 ± 0.6	0.0912 ± 0.004	0.2566 ± 0.05	2.7
		318.15	166.9 ± 0.7	0.0969 ± 0.005	0.3303 ± 0.06	3.9
39.96 % C_2H_5OH	0.050 - 2.000	288.15	203.2 ± 1	0.1109 ± 0.010	0.1165 ± 0.12	4.6
		298.15	205.1 ± 1	0.1204 ± 0.010	0.1714 ± 0.13	4.9
		318.15	212.6 ± 1	0.1189 ± 0.010	0.3497 ± 0.12	4.9
9.82 % *1*-C_3H_7OH	0.0485 - 3.002	298.15	151.9 ± 0.4	0.0877 ± 0.003	0.2352 ± 0.03	2.0
		318.15	159.6 ± 0.3	0.0955 ± 0.002	0.2432 ± 0.03	2.0
19.7 % *1*-C_3H_7OH	0.051 - 1.500	298.15	171.6 ± 0.2	0.0818 ± 0.003	0.2818 ± 0.02	1.0
		318.15	178.3 ± 0.4	0.0906 ± 0.006	0.3442 ± 0.04	2.0
29.62 % *1*-C_3H_7OH	0.049 - 1.199	298.15	190.5 ± 0.2	0.0955 ± 0.007	0.2703 ± 0.04	1.0
		318.15	197.4 ± 0.4	0.1073 ± 0.01	0.3955 ± 0.06	2.0
39.56 % *1*-C_3H_7OH	0.051 - 0.850	298.15	205.9 ± 0.3	0.0911 ± 0.01	0.5852 ± 0.04	1.0
		318.15	216.4 ± 0.5	0.1120 ± 0.02	0.6472 ± 0.09	2.0
10.0% *iso*-C_3H_7OH	0.050 - 3.000	298.15	151.4 ± 0.5	0.0810±0.02	0.3417 ± 0.09	3.2
20.0% *iso*-C_3H_7OH	0.050 - 2.500	298.15	179.9 ± 0.5	0.0859 ± 0.004	0.3290 ± 0.04	2.0
30.0% *iso*-C_3H_7OH	0.050 - 2.000	298.15	198.0 ± 0.3	0.0884 ± 0.004	0.4658 ± 0.03	1.4
40.0 % *iso*-C_3H_7OH	0.100 - 1.400	298.15	222.7 ± 0.5	0.0943 ± 0.007	0.6030 ± 0.06	1.0
49.9 % *iso*-C_3H_7OH	0.050 - 0.952	298.15	243.6 ± 0.6	0.0939 ± 0.02	1.0109 ± 0.09	2.0
58.5% *iso*-C_3H_7OH	0.050 - 0.710	298.15	270.5 ± 0.3	0.1266 ± 0.01	1.2882 ± 0.05	1.0
3.00 % *1*-C_4H_9OH	0.051 - 2.006	288.15	118.4 ± 0.4	0.0774 ± 0.004	0.2090 ± 0.04	1.6
		298.15	120.1 ± 0.4	0.0808 ± 0.004	0.2999 ± 0.03	1.5
		318.15	127.7 ± 0.6	0.0881 ± 0.006	0.3200 ± 0.05	2.5
4.49 % *1*-C_4H_9OH	0.049 – 1.248	288.15	121.4 ± 0.6	0.0763 ± 0.01	0.2032 ± 0.08	2.5
		298.15	123.8 ± 0.6	0.0939 ± 0.01	0.2246 ± 0.08	2.3
		318.15	130.8 ± 0.5	0.0903 ± 0.01	0.3146 ± 0.07	1.8
5.66 % *1*-C_4H_9OH	0.050 – 0.601	288.15	124.0 ± 0.4	0.0952 ± 0.04	0.1501 ± 0.15	2.0
		298.15	127.0 ± 0.4	0.1257 ± 0.03	0.1284 ± 0.11	1.4
		318.15	134.1 ± 0.8	0.0883 ± 0.06	0.3314 ± 0.22	2.3
3.00 % *iso*-C_4H_9OH	0.051 - 2.500	288.15	119.0 ± 0.6	0.0796 ± 0.006	0.1269 ± 0.06	2.8
		298.15	121.3 ± 0.5	0.0839 ± 0.005	0.2565 ± 0.05	2.6
		318.15	128.1 ± 0.8	0.0905 ± 0.007	0.3782 ± 0.08	4.2

w_{Alc}, wt. %	m, mol·kg^{-1}	T, K	$-E_0\times 10^3$, V	$\beta^{(0)}_{NaCl}$, kg·mol^{-1}	$\beta^{(1)}_{NaCl}$, kg·mol^{-1}	$s_0(E)\times 10^4$
1	2	3	4	5	6	7
4.50 % *iso*-C_4H_9OH	0.049 – 1.799	288.15	1209 ± 0.3	0.0877 ± 0.004	0.1215 ± 0.03	1.2
		298.15	1242 ± 0.3	0.0917 ± 0.004	0.1976 ± 0.03	1.4
		318.15	1301 ± 0.5	0.0873 ± 0.007	0.3961 ± 0.05	2.0
5.66 % *iso*-C_4H_9OH	0.050 – 0.600	288.15	124.8 ± 0.3	0.0765 ± 0.02	0.2322 ± 0.07	0.9
		298.15	127.0 ± 0.3	0.0685 ± 0.02	0.3696 ± 0.07	0.9
		318.15	134.2 ± 0.5	0.0179 ± 0.03	0.6969 ± 0.12	1.6
2.00 % *1*-$C_5H_{11}OH$	0.050-0.600	288.15	114.7 ± 0.6	0.0376 ± 0.04	0.3142 ± 0.15	1.9
		298.15	117.6 ± 0.6	0.0762 ± 0.03	0.2634 ± 0.13	1.8
2.00 % *iso*-$C_5H_{11}OH$	0.050-0.650	288.15	114.8 ± 0.4	0.0598 ± 0.02	0.2772 ± 0.08	1.1
		298.15	117.3 ± 0.5	0.0768 ± 0.03	0.3012 ± 0.10	1.6

Table 3. Pitzer parameters for solutions of sodium chloride in water-alcohol solvent.

As an example, the variation of mean ionic activity coefficient versus the electrolyte molality and various mass fraction percents of ethanol-water mixed solvents at 298.15 K, are shown in Fig. 2 *a*. The calculated values of $\gamma_{\pm}$(NaCl) in mixed *1*-propanol-water solvent are represented by solid lines in Figs. 2 *b*. The transparent symbols correspond to the dependence of mean activity coefficient of the alcohol-free solution obtained in the present study. All the curves show a typical profile of the variation of $\gamma_{\pm}$ with concentrations that, as is well known, are governed by two types of interactions: ion–ion and ion–solvent. For a given temperature, the minimum value of mean ionic activity decreases with the increase of wt.% of alcohol. The trend is identical at other temperatures.

If the parameters of the thermodynamic model are defined the calculation of any thermodynamic function is a routine mathematical operation. For example, Eq. (6) may be used for the estimation of the osmotic coefficient. The concentration dependences of this function in the ternary solutions containing 19.98 wt.% of C_2H_5OH and 3 wt.% of *iso*-C_4H_9OH are shown in Fig. 3.

The NaCl-H_2O-C_2H_5OH solutions belong to the most investigated system in comparison with other analogical objects, so it's possible to estimate the quality of our experimental data not only for aqueous solvent but for mixed solvent as well. Fig.2 *a* demonstrates a good agreement between literature data and the results of the present investigation. Hereinafter the system with ethanol as organic component was accounted as an object to test the various approaches to data approximation. The aim was to reveal the main factors that affected the accuracy of partial and integral properties determination based on the EMF measurements with ion-selective electrodes. The next factors were investigated: (a) a number of experimental points included in approximation; (b) type of thermodynamic model.

Correlation between the quantity of experimental data and the results of calculation of mean ionic activity coefficients in the NaCl-H_2O-C_2H_5OH system at 298.15 K and w_{Alc} = 9.99 % is illustrated by Table 4. The Pitzer's model parameters at various numbers of input data (EMF, m_{NaCl}) are listed. An analysis of the data presented in Table 4 shows that an increase in the number of experimental data (i.e. expansion of concentration range) does give noticeable advantages for the description of Pitzer's model. The standard deviation of approximation for more than seven E,m-pairs varies slightly (from 2.4 to 2.8) unlike the significance of parameters; the approximation with only seven experimental points gave insignificant values. To improve a quality of description it is necessary to use more than 10 primary (*E, m*) measurements. In many cases it is impossible to extend the concentration range due to the restrictions for lower and upper limit of molality. So, a number of 12-14 points were considered as optimum set of data in each series of experiments. Other thermodynamic models (Pitzer-Simonson model or eNRTL) are less sensitive to the amount of input data. The advantage of those models is ability to use entire set of experimental data unlike Pitzer model where only the results for solution with a constant water/alcohol ratio are accounted.

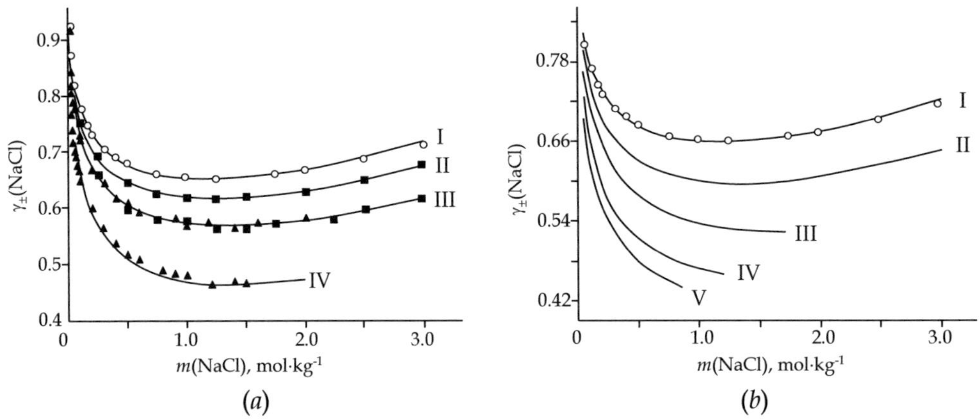

Fig. 2. Mean ionic activity coefficients of NaCl in a NaCl-H_2O-C_2H_5OH (*a*) and NaCl-H_2O-*1*-C_3H_7OH (*b*) solution at 298.15 K. Transparent symbols (○) correspond to the NaCl-H_2O solution without alcohol (Lide, 2007-2008; Silvester & Pitzer, 1977), fill symbols are the literature data (▲ - Esteso et al., 1989; ■ - Lopes et al., 2001), lines are the result of present investigation. Number I, II, III, IV and V correspond to the experimental data for solutions containing (*a*) 0, 9.99, 19.98 and 39.96 wt. % of the ethanol; (*b*) 0, 9.82, 19.70, 29.62, 39.96 wt. % 1-propanol, respectively.

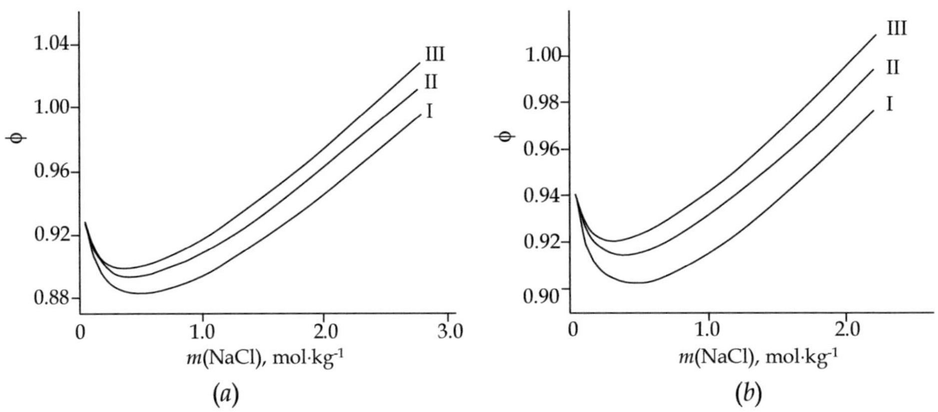

Fig. 3. Osmotic coefficients in NaCl-H_2O-C_2H_5OH (w_{Alc}= 19.98%) (*a*) and NaCl-H_2O-*iso*-C_4H_9OH (w_{Alc}= 3.00%) (*b*) solutions, where the numbers I, II, and III indicate the experimental data at 288.15, 298.15, and 318.15 K, respectively.

Number of experimental points	$-E_0\times 10^3$, V	$\beta^{(0)}_{NaCl}$, kg·mol^{-1}	$\beta^{(1)}_{NaCl}$, kg·mol^{-1}	$s_0(E)\times 10^4$
7	138.6 ± 1.7	0.1074 ± 0.132	0.2132 ± 0.46	2.8
8	138.3 ± 1.1	0.0810 ± 0.053	0.2994 ± 0.23	2.6
9	138.3 ± 0.9	0.0809 ± 0.027	0.2998 ± 0.14	2.5
10	138.2 ± 0.7	0.0770 ± 0.017	0.3162 ± 0.10	2.4
11	138.4 ± 0.6	0.0823 ± 0.009	0.2889 ± 0.07	2.4
12	138.4 ± 0.5	0.0842 ± 0.007	0.2785 ± 0.06	2.4
13	138.5 ± 0.5	0.0850 ± 0.004	0.2728 ± 0.05	2.3
14	138.6 ± 0.6	0.0884 ± 0.004	0.2468 ± 0.05	2.8

Table 4. A comparison of Pitzer's model parameters with various set of input EMF data. The $NaCl-H_2O-C_2H_5OH$ system (9.99 wt.%) at 298.15 K.

For the estimation of model's type influence on the results of calculation the Pitzer theory was compared with the more complex Pitzer-Simonson model. According to the Pitzer formalism, the mean ionic activity coefficient of a strong electrolyte is defined within the molality mode related to the "ideal dilute standard". But the definitions which use mole fractions are more suitable or even necessary for the calculations of phase equilibria. The Pitzer-Simonson model is based on mole fractions. So initial values of $\gamma_{\pm}$(NaCl) were at first converted to the mole fraction basis by the equation

$$\gamma_x = \gamma_m(1+0.001mM_s) = \left(\gamma_{\pm}\right)^2 m(1+0.001mM_s), \tag{15}$$

where M_s – is the mean molar mass of solvent. Calculated activity coefficients γ_x were compared with those obtained by Eq. (13). This equation includes two groups of parameters for the description of the Gibbs excess energy of mixing in the frame of Pitzer-Simonson formalism: three pairs of binary parameters "salt-water", "alcohol-water", "salt-alcohol" and one ternary parameter. The long-range parameter for sodium chloride in diluted range (B_{MX}) and short-range interaction parameters (W_{1NaCl} and U_{1NaCl}) related to salt (NaCl) and water (1) were estimated on the base of literature data (Pitzer & Mayorga, 1973) and own experiments. Results of calculation are presented in the Table 5. The binary interaction parameters for a system containing water (1) and ethanol (2) were determined by the minimization of the objective function, which was the sum of the squares of the relative deviations between the calculated and experimental values for different data sets in the $H_2O-C_2H_5OH$ system (see, for instance, Shishin et al. 2010; Konstantinova et al., 2011) or by using literature data (Lopes et al., 2001). The values of short-range interactions parameters W_{nMX}, V_{nMX}, U_{nMX} (n = 1,2) related to sodium chloride and component of mixed solvent, given in Table 5, were obtained from the results of EMF measurements of electrochemical cell (I).

The differences between the activity coefficients of NaCl obtained from the two models are shown in Fig.4 for various temperatures and solvent compositions. The results demonstrate that both methods are adequate for treating the EMF measurements. A comparison of calculations with two models gives the largest difference of 5 % for the alcohol rich solution. At the same time the mean ionic activity coefficients of NaCl are consistent with the published data (Lopes et al., 2001) if the uniform Pitzer-Simonson model is used for data approximation; maximum error is equal to 1.5 % for the solution with w_{Alc} = 19.98% at 318.15

K. An analysis of the data presented in Table 3 and Table 5 confirms the conclusion of Lopes that the variation of the activity coefficients of salt with the temperature or with the alcohol content is always smoother when the Pitzer–Simonson equations are used.

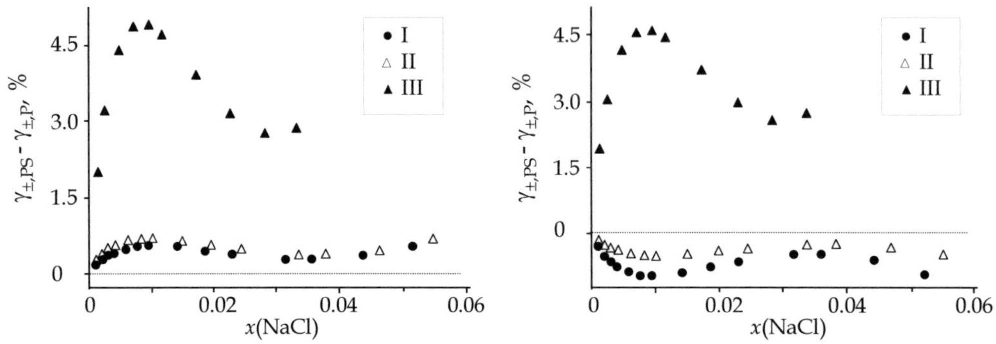

Fig. 4. The difference between Pitzer-Simonson and Pitzer model calculations at 288.15 K (*a*) and 298.15 K (*b*). A number I, II and III corresponds to various composition of solvent: 9.99, 19.98 and 39.96 w.%, respectively.

T, K	288.15	298.15	318.15
w_{12}	1.1934	1.2051	1.2285
u_{12}	0.2123	0.2583	0.3505
$-W_{1NaCl}$	6.1366	5.9187	5.6356
$-U_{1NaCl}$	4.5747	4.1214	3.5427
B_{NaCl}	12.6266	12.8566	14.2752
Z_{12NaCl}	7.2780	2.0676	3.0343
$-W_{2NaCl}$	35.1662	24.7911	16.9476
$-U_{2NaCl}$	31.0909	27.0080	15.8373
$-E^0(w_{Alc}$-9.99%$)\times10^3$,V	373.67	373.76	377.08
$-E^0(w_{Alc}$-19.98%$)\times10^3$, V	439.95	428.07	411.08
$-E^0(w_{Alc}$-39.96%·$)\times10^3$, V	600.63	552.08	488.78
$s_0(E)\times10^3$	4.2	4.4	4.4

Table 5. Interaction parameters of the Pitzer-Simonson model for the binary NaCl-H_2O and ternary NaCl-H_2O-C_2H_5OH solutions.

Unlike in the Pitzer formalism, some of the binary interaction parameters in the Pitzer-Simonson model are defined independently by the approximation experimental data in Water-Alcohol and Water-Salt systems. In principle, the difference in those parameters can cause discrepancies in the results of calculation from various scientific groups. For the estimation of possible error we calculate the mean ionic activity coefficients with a various sets of water-ethanol interaction parameters. Model parameters were calculated using the primary experimental data about vapour-liquid equilibria that were obtained by various authors in the H_2O-C_2H_5OH system. It was revealed that this factor does not affect to the result of γ_x calculation; the difference in values is less than a hundredth of a percent.

5.2 Integral properties of ternary solutions

The integral properties of solutions were calculated with the Pitzer, Pitzer-Simonson model and by the Darken method. In the former case, the molality concentration scale was used, and the mixed solvent and the extremely dilute solution of the salt in this solvent were taken as the base. The formulas used for the calculation of the integral functions are listed in Section 3.

One of the features of the EMF-experiment with ion-selective electrodes is a step-by-step increase of the molality at a fixed ratio of the solvent components in every experiment. As a result, the composition of the system under investigation is changing through the quasibinary section of ternary system. This allows to use the Darken method for the calculation of the integral properties of such solutions. The essence of this procedure is an integration of the Gibbs-Duhem equation in the ternary solution at a constant ratio for any two components. This method is widely used for the results of the electrochemical measurements for ternary metal alloys but has not been applied yet for the aqueous electrolyte systems.

Since solutions from one series of the EMF-measurements belong to particular NaCl-$[H_2O+Alc]_{H_2O/Alc}$ section of the ternary system with fixed relation of alcohol and water, the following equation can be used for the excess Gibbs energy of such solutions:

$$G_x^{ex} = (1 - x_{\text{NaCl}})\left[G_{x_{\text{NaCL}}=0}^{ex} + \int_0^{x_{\text{NaCl}}} \frac{\mu_{\text{NaCl}}^{\text{ex}}}{(1\text{-}x_{\text{NaCl}})^2} dx_{\text{NaCl}} \right]_{\frac{x_{\text{Alc}}}{x_{\text{H}_2\text{O}}}} . \quad (16)$$

where G_x^{ex} and $G_{x_{\text{NaCL}}=0}^{ex}$ are the excess integral Gibbs energy of solution at mole fractions of sodium chloride equal to x_{NaCl} or 0, respectively. The excess chemical potential of NaCl is connected with the activity coefficient in various concentration scales by the relation

$$\mu_{\text{NaCl}}^{\text{ex}} = RT\ln\gamma_x = RT\left(2\ln\gamma_\pm + \ln\left(m\left(1+0.001m\left\{\left(1-x_{\text{Alc}}^\infty\right)M_{\text{w}} + x_{\text{Alc}}^\infty M_{\text{Alc}}\right\}\right)\right)\right), \quad (17)$$

where M_{w}, M_{Alc} are the molar masses of water and alcohol, and x_{Alc}^∞ is the mole fraction of alcohol in salt-free solvent. The temperature-concentrations dependences of the mean ionic activity coefficient of NaCl and the excess Gibbs energies of water-alcohol binary solutions ($G_{x_{\text{NaCl}=0}}^{ex}$) are necessary for the calculation of integral properties of ternary solutions according to Eq. (16). The former may be obtained from the approximation of the results of the EMF measurements. In the case of ISE two unknown values - E_0 and γ - are presented in Eq.(14) unlike common galvanic cells with one unknown γ. The Darken method has obvious advantages for the processing of electrochemical experiments with common electrodes due to the possibility to provide a numerical integration of the Gibbs-Duhem equation. The result of this calculation does not depend upon the type of the model; thus the error in integral properties determination may be reduced. The approximation of EMF-ISE measurements includes a choice of a thermodynamic model. In such situation Darken method can be considered as an alternative way to calculate Gibbs energy of solution in molar mode. For instance, if the Pitzer formalism is used then the excess Gibbs energy of the ternary solution is calculated by the equation (for more detail see Mamontov et al., 2010):

$$G_x^{ex} = \frac{G_{x_{\mathrm{NaCL}}=0}^{ex}}{(1+0.001\cdot mM)} + \frac{2RT\cdot M}{(1000+mM)}\left(\begin{array}{l} -\frac{2}{1.2}A_\phi m\ln(1+1.2m^{1/2}) + \beta_{\mathrm{NaCl}}^{(0)}\cdot m^2 + \frac{\beta_{\mathrm{NaCl}}^{(1)}m}{2}(1-(1+2m^{1/2})\exp(-2m^{1/2})) + \\ +\frac{C_{NaCl}^{\gamma}}{2}m^3 + \frac{1}{2}\left(m\cdot\ln(m) - 2m + \frac{1000}{M}(1+0.001mM)\ln(1+0.001mM)\right) \end{array}\right) \quad (18)$$

To calculate the excess Gibbs energies of water-alcohol binary solutions it is necessary to construct a thermodynamic model of those solutions. The model parameters of aqueous solutions of alcohols were proposed in numerous studies (see, for example, Gmehling & Onken, 1977 or subsequent Editions). In the present study we used a results of own assessment of Water-Alcohol systems with NRTL model (Shishin et al., 2010; Konstantinova et al., 2011). The main merit of our approach is a use of unified equation of the Gibbs energy of a water-alcohol solution for the description of various types of equilibria: vapor-liquid and liquid-liquid.

Pure alcohol and water, along with a hypothetical infinitely dilute solution of NaCl in the mixed solvent, were chosen as the standard state of the solution components during integration by the Darken's method. To go to the unified reference system (pure solvents and the extremely dilute solution of the salt in water), the so-called Born contribution accounting for the change in the Gibbs energy ($G_{\mathrm{Born}}^{\mathrm{ex}}$) was used (see Eq.(4)). The Gibbs energy of formation of the solution in this reference system can be written as follows:

$$\Delta_{mix}G = RT\cdot\left(x_{\mathrm{H_2O}}\ln x_{\mathrm{H_2O}} + x_{\mathrm{Alc}}\ln x_{\mathrm{Alc}} + x_{\mathrm{NaCl}}\ln x_{\mathrm{NaCl}}\right) + G_x^{\mathrm{ex}} + G_{\mathrm{Born}}^{\mathrm{ex}}\,. \quad (19)$$

Other thermodynamic functions were obtained from Eq.(19) by means of common thermodynamic relations, for example, Gibbs-Helmholtz equation. Next figures illustrate some results of the calculation. Fig. 5 *a,b* shows the isothermal (298.15 K) sections of Gibbs energy surface for the solution formed by sodium chloride, water, and *iso*-propanol with a constant ratio of water and alcohol in solvent, and with a constant mole fraction of sodium chloride. Since all electrochemical measurements were carried out with the use of homogeneous mixtures, no anomalies are observed in the curves in Fig. 5 in the region of the measured compositions (solid lines), and they are shifted with respect to each other in the expected way. Fig. 6 shows the sections of the surface of the enthalpy of mixing of the solution with *iso*-butanol at a constant composition of the solvent and a constant fraction of NaCl at 298.15 K. The relative enthalpy of the ternary solution changes only slightly in the temperature range under consideration.

The Darken method is theoretically justified, so there is no reason to doubt the correctness of the results. Additionally we compared the values of thermodynamic function been calculated with this approach and Pitzer-Simonson model. As can be seen from the Fig.7 the results are in excellent agreement.

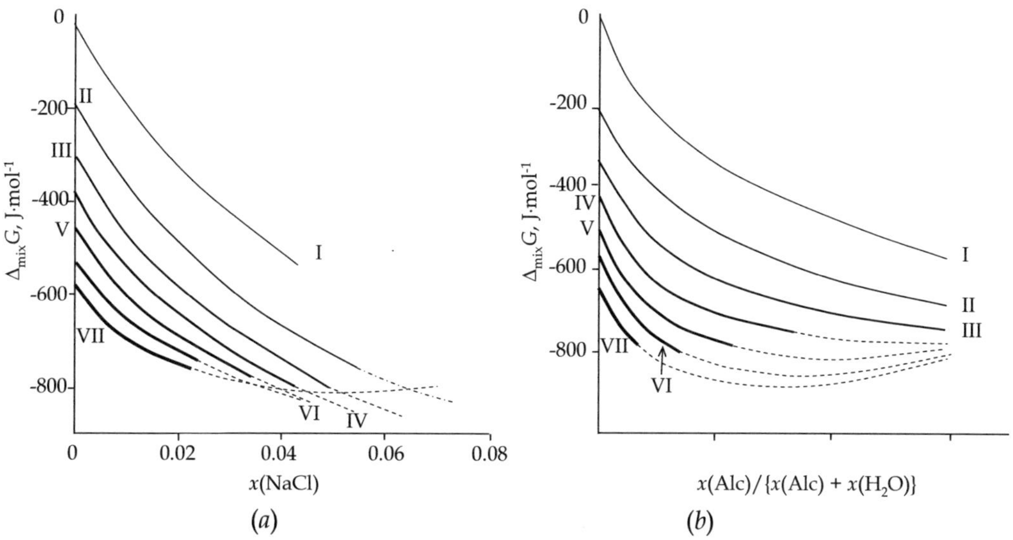

Fig. 5. The isothermal (298.15 K) sections of Gibbs energy surface of solutions of H_2O–*iso*-C_3H_7OH–NaCl with constant (*a*) the ratio of water and alcohol; (*b*) the weight fraction of alcohol in solvent. The numbers correspond to solutions with the weight fraction of alcohol in solvent is: (I) 0, (II) 10.0, (III) 20.0, (IV) 30.0, (V) 40.0, (VI) 49.9, (VII) 58.5 wt. % (*a*); and with mole fraction of sodium chloride equal to: (I) 0, (II) 0.01, (III) 0.02, (IV) 0.03, (V) 0.04, (VI) 0.05, (VI) 0.06 (*b*). Dashed lines indicate the extrapolated values of the $\Delta_{mix}G$ function outside the range of experimental research. The line width increases with the increasing of (*a*) the weight fraction of alcohol in solvent; (*b*) the mole fraction of NaCl.

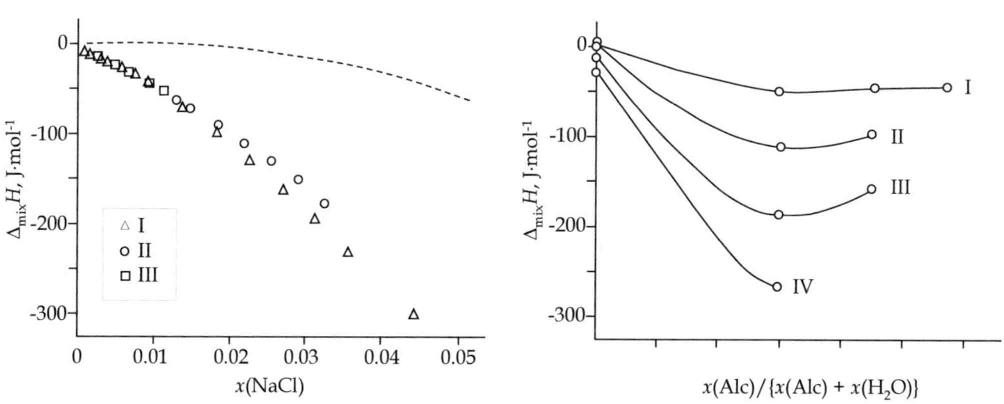

Fig. 6. Isothermal sections of the surface of the enthalpy of mixing of the NaCl–H_2O–*iso*-C_4H_9OH solution at 298.15 K along the sections with constant water-to-alcohol ratios (*a*); the weight fraction of the alcohol in the solvent (w_{Alc}) is (I) - 3.0; (II) - 4.5; (III) - 5.66 wt. %, dashed line – without alcohol; (*b*) along the sections with constant mole fractions of sodium chloride; x_{NaCl} is (I) - 0.01; (II) - 0.02; (III) - 0.03; (IV) - 0.04. The points in each section at (*b*) are connected by a solid line.

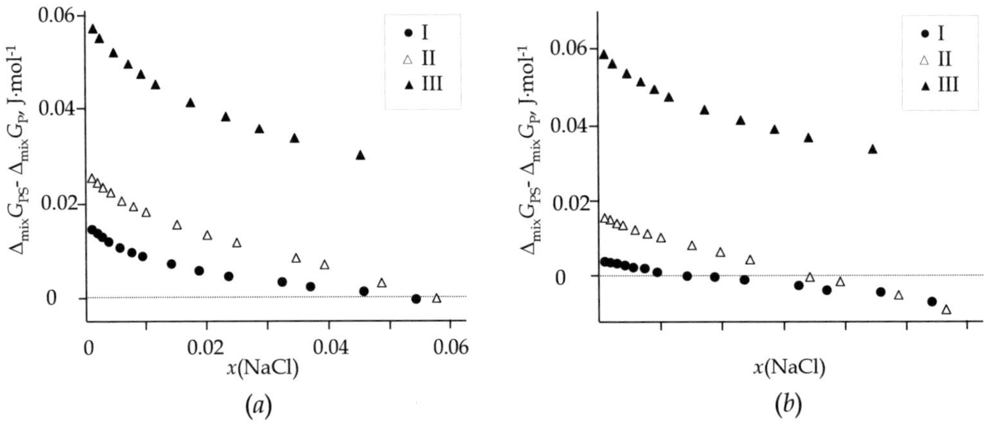

Fig. 7. The difference between results of Gibbs energy calculation of the $NaCl\text{-}H_2O\text{-}C_2H_5OH$ solution at 298.15 K (*a*) ad 318.15 K (*b*) with Darken method and Pitzer-Simonson model. A number I, II, III corresponds to various composition of solvent: 9.99, 19.98 and 39.96 w.%, respectively.

5.3 Standard thermodynamic functions of transfer of salt from water to the mixed solvent

The electrochemical measurements allowed us not only to calculate the mean ionic activity coefficients, the Gibbs energies, and the enthalpies of mixing of ternary solutions but also to estimate the standard thermodynamic functions of transfer of salt from water to a mixed solvent. The standard Gibbs energy of transfer is defined as the difference between the standard Gibbs energy per mole of electrolyte in a pure solvent, usually water, and that in another pure or mixed solvent. It is a measure of the change in the total energy of the solute when it is transferred from one solvent to another at infinite dilution and can be easily calculated according to the expression:

$$\Delta_{\text{tr}} G_T^{\text{o}} = -F(E_{0,\text{s}} - E_{0,\text{w}}) \tag{20}$$

where $E_{0,s}$ and $E_{0,w}$ are the standard electrode potentials of the electrochemical cell (I) for the mixed solvent and water, respectively. The values $\Delta_{\text{tr}} G_T^{\text{o}}$ were calculated for each temperature and the composition of the solution based on the data presented in Table 6. Fig. 8 *a* shows $\Delta_{tr}G$ *vs.* n (number of carbon atoms in *n*- and *iso*-alcohol) for ternary solutions with w_{Alc} = 5 % at 298 K. It is surprising the decrease of Gibbs energy of transfer for aqueous-alcohol mixtures with n > 3. Apparently, this phenomenon requires further investigation.

w_{alc}, %	$-(E_{0,s} - E_{0,w})$, mV			$\Delta_{tr}G^{\circ}_{T}$, J·mol^{-1}		
	288.15 K	298.15 K	318.15 K	288.15 K	298.15 K	318.15 K
Water-ethanol solvent						
9.99	22.4 ± 0.4	22.4 ± 0.4	22.5 ± 0.6	2.16 ± 0.04	2170 ± 60 1910 ± 100 (Mazzarese & Popovych, 1983 5600 ± 100 (Kalida et al., 2000)	2140 ± 80
19.98	44.6 ± 0.5	44.6 ± 0.6	43.6 ± 0.7	4.30 ± 0.05	4300 ± 60 3880 ± 100 (Mazzarese & Popovych, 1983) 6400 ± 100 (Kalida et al., 2000)	4210 ± 70
39.96	89.5 ± 1.0	89.0 ± 1.0	89.3 ± 1.0	8.64 ± 0.10	8590 ± 100 (*) 8330 ± 100 (Mazzarese & Popovych, 1983) 8000 ± 100 (Kalida et al., 2000)	8620 ± 100
Water-*n*-propanol solvent						
9.82	-	22.4 ± 0.7	23.2 ± 0.8	-	2161 ± 70 2036 ± 80 (Gregorowicz et al., 1996)	2238 ± 80
19.70	-	42.1 ± 0.6	41.9 ± 0.7	-	4062 ± 60 3908 ± 80 (Gregorowicz et al., 1996)	4042 ± 70
29.62	-	61.0 ± 0.6	61.0 ± 0.7	-	5885 ± 60 5649 ± 80 (Gregorowicz et al., 1996)	5885 ± 70
39.56	-	76.4 ± 0.7	80.0 ± 0.8	-	7371 ± 70 7451 ± 80 (Gregorowicz et al., 1996)	7718 ± 80
Water-*n*-butanol solvent						
3.00	-4.7 ± 0.4	-4.0 ± 0.4	-4.4 ± 0.6	450 ± 40	390 ± 40 640 ± 70 (Chu et al., 1987)	430 ± 60
4.49	-7.7 ± 0.6	-7.7 ± 0.6	-7.5 ± 0.5	740 ± 60	740 ± 58 960 ± 100 (Chu et al., 1987)	720 ± 50
5.66	-10.3 ± 0.4	-10.9 ± 0.4	-10.8 ± 0.8	990 ± 40	1050 ± 40 1200 ± 120 (Chu et al., 1987)	1040 ± 80
Water-*n*-pentanol solvent						
2.00	-1 ± 0.6	-1.5 ± 0.5	-	99 ± 58	147 ± 48	-

Table 6. Standard Gibbs energy of transfer of NaCl from H_2O to the H_2O (100-w_{alc}%) + $C_nH_{2n+1}OH$ (w_{alc}%) mixed solvent at various temperatures.

It can be seen that the values of the standard functions of transfer determined in the present study do not contradict to those recommended in the literature. Data for the propanols are in satisfactory agreement with results (Gregorowicz et al., 1996). In the case of butanols the values of the standard functions of transfer at 298.15 K are shifted to smaller values compared to those recommended in the study (Chu et al., 1987). In all cases an increase in the standard Gibbs energy of transfer is observed with the increase of alcohol in the mixture, which would indicate a decrease in hydration of electrolyte in the mixture. Using the Feakins and French equation (Feakins & French, 1957), it is possible to estimate the primary hydration number of the electrolyte based on the dependency between the standard electromotive force of the cell and the logarithm of the mass fraction of water in the mixture:

$$\Delta E_0 = E_{0,s} - E_{0,w} = N_h k_B \log w \quad (21)$$

where k_B is a is the Boltzmann constant. This value varies from 8 for methanol to 0.5 for butanol and coincides for different homologs of alcohols within error. Fig. 8 *b* shows the plot of N_{hyd} *vs.* n. The solid lines in both graphics have no physical meaning, they represent a trend.

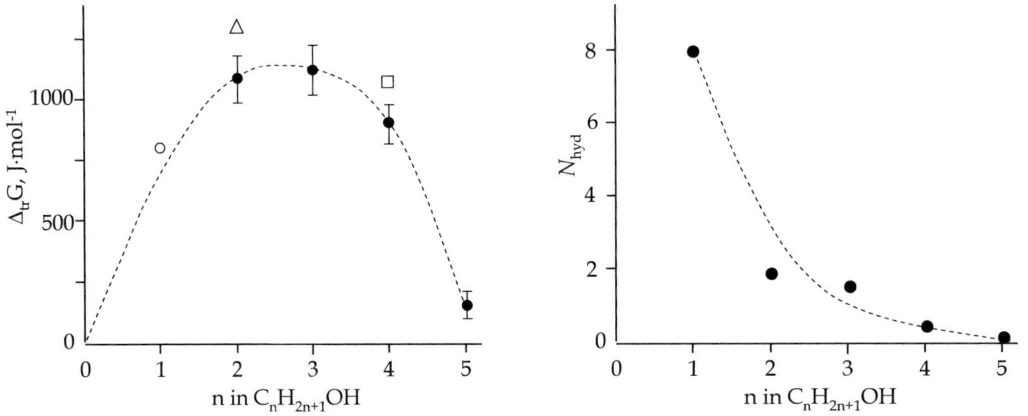

Fig. 8. Plot of $\Delta_{tr}G$ *vs.* n (in $C_nH_{2n+1}OH$) (*a*) and N_{hyd} *vs.* n (*b*) ; the filled symbol - present work, the transparent symbol (Mazzarese & Popovych, 1983; Chu et al., 1987). T = 298.15 K

6. Conclusion

Systematic studies of electrochemical cells with Na+ and Cl- ion-selective electrodes in aqueous-alcohol solvents were performed. Analysis of the results have shown that the major sources of uncertainty of the partial properties in the EMF measurements with the ISE are the choice of model solution and the amount of input data that take into account at the approximation. Therefore, when comparing the activity coefficients of electrolyte obtained in various researches groups should pay particular attention to the method of data processing.

The Darken method was first used to calculate the integral properties of electrolyte solutions. It was shown that the Gibbs energy of mixing calculated by the Pitzer-Simonson model and the Darken method are in excellent agreement.

The results of present investigation can be used to the verification of proposed thermodynamic models and for the development of alternative theories of electrolyte solutions.

7. Acknowledgment

This work was supported by the RFBR (project 09-03-01066) and the URALCHEM holding.

8. References

Achard, C.; Dussap, C.G. & Gros, J.B. (1994). Representation of vapour-liquid equilibria in water-alcohol-electrolyte mixtures with a modified UNIFAC group-contribution method. *Fluid Phase Equilibria*, Vol. 98, pp. 71-89, ISSN 0378-3812

Balaban, A.; Kuranov, G. & Smirnova, N. (2002). Phase equilibria modeling in aqueous systems containing 2-propanol and calcium chloride or/and magnesium chloride. *Fluid Phase Equilibria*, Vol. 194-197, pp. 717-728, ISSN 0378-3812

Bald, A.; Gregorowicz, J. & Szergis, A. (1993). Thermodynamic and conductomertic studies on NaI solutions in water-isobutanol mixtures at 298.15K. *Phys. Chem. Liq.*, Vol. 26, pp. 121-133, ISSN 0031-9104

Bratov, A.; Abramova, N. & Ipatov, A . (2010). Recent trends in potentiometric sensor arrays—A review. *Analytica Chimica Acta*, Vol. 678, pp. 149–159, ISSN 0003-2670

Buck, R. P. & Lindner, E. (2001) Tracing the History of Selective Ion Sensors. Progress in ISE development has occurred rapidly in the past 35 years, with promising innovations still on the horizon. *Analytical Chemistry*, Vol. 73, No 3, pp. 88-97, ISSN 0003-2700

Chen, C.C.; Britt, H.I.; Boston, & Evans, J.F. (1982). Local composition model for excess Gibbs energy of electrolyte systems. Part I: single solvent, single completely dissociated electrolyte systems. *AIChE Journal*, Vol. 28, No. 4, pp. 588-596, ISSN 1547-5905

Chen, C.C.; Mathias, P.M. & Orbey, H. (1999) Use of hydration and dissociation chemistries with the electrolyte-NRTL model. *AIChE Journal*, Vol. 45, pp. 1576–1586, ISSN 1547-5905

Chen, C. C.; Bokis, C. P. & Mathias, P. (2001). Segment-based excess Gibbs energy model for aqueous organic electrolytes. *AIChE Journal*, Vol. 47, pp. 2593-2602, ISSN 1547-5905

Chen, C.-C. & Mathias, P. M. (2002). Applied Thermodynamics for Process Modeling. *AIChE Journal*, Vol. 48, No.2, pp.194-200, ISSN 1547-5905

Chen, C.C. & Song, Y. (2004). Generalized electrolyte-NRTL model for mixed-solvent electrolyte systems. *AIChE Journal*, Vol. 50, No. 8, pp. 1928-1941, ISSN 1547-5905

Chen, C.C. (2006). Toward development of activity coefficient models for process and product design of complex chemical systems. *Fluid Phase Equilibria*, Vol.241, pp.103-112, ISSN 0378-3812

Chu, D.Y.; Zhang, Q. & Liu, R.L. (1987). Standard Gibbs free energies of transfer of NaCl and KCl from water to mixtures of the four isomers of butyl alcohol with water. *J. Chem. Soc., Faraday Trans.*, Vol. 83, pp. 635-644, ISSN 0956-5000

Clegg, S.L. & Pitzer, K.S. (1992). Thermodynamics of multicomponent, miscible, ionic solutions : generalized equations for symmetrical electrolytes. *Journal of Physical Chemistry*, Vol. 96, No. 23, pp. 3513-3520, ISSN 1520-6106

Deyhimi, F.; Karimzadeh, Z. & Abedi, M. (2009). Pitzer and Pitzer-Simonson-Clegg modeling approaches: Ternary HCl + methanol + water electrolyte system. *J. Molecular Chem.* Vol.150, pp. 62-67, ISSN 0167-7322

Deyhimi, F. & Karimzadeh, Z. (2010). Pitzer and Pitzer-Simonson-Clegg modeling approaches: Ternary HCl + 2-Propanol + water electrolyte system, *J. Solution Chemistry*, Vol.39, pp. 245–257, ISSN 0095-9782

Esteso, M.A.; Gonzalez-Diaz, O.M.; Hernandez-Luis, F.F. & Fernandez-Merida L. (1989). Activity Coefficients for NaCl in Ethanol-Water Mixtures at 25 C. *J. Solution Chemistry*, Vol. 18, No. 3, pp.277-288, ISSN 0095-9782

Feakins, D. & French, C.M. (1957). E.m.f. measurements in ethyl methyl ketone-water mixtures with the cell H_2(Pt)|HCl|AgCl-Ag. With an appendix on triethylene glycol-water systems. *J. Chem. Soc.*, pp. 2284-2287

Frenkel, M.; Hong, X.; Wilhoit, R.C. & Hall, K.R. (1998). *Thermodynamic Properties of Organic Compounds and their Mixtures Springer*, ISBN 3-540-66233-2, Berlin, Germany

Ganjali, M.R.; Norouzi, P.; Faridbod, F.; Rezapour, M. & Pourjavid M.R. (2007). One Decade of Research on Ion-Selective Electrodes in Iran (1996-2006) *J. Iran. Chem. Soc.*, Vol. 4, No. 1, pp. 1-29, ISSN 1735-207X

Gmehling, J. & Onken, U. (1977). *Vapour-liquid equilibrium data collection. Aqueous-organic systems.* Dechema Chemistry Data Series. Part1. Vol.1, ISBN 3-921567-01-7, Germany

Gregorowicz, J.; Szejgis, A. & Bald,A. (1996). Gibbs energy and conductivity properties of NaCl solutions in water-*iso*-propanol mixtures at 298.15K. *Phys. Chem. Liq.*, Vol.32, pp. 133-142, ISSN 0031-9104

Ipser, H.; Mikula, A. & Katayama, I. (2010). Overview: The emf method as a source of experimental thermodynamic data. *CALPHAD*, Vol. 34, pp. 271-278, ISSN 0364-5916

Kalida, C.; Hefter, G. & Marcus, Y. (2000). Gibbs energies of transfer of cations from water to mixed aqueous organic solvents. *Chemical Reviews*, Vol. 100, No.3, pp. 821-852, ISSN 0009-2665

Kikic, I.; Fermeglia, M. & Rasmussen, P. (1991). UNIFAC prediction of vapour-liquid equilibria in mixed solvent-salt systems. *Chem. Eng. Sci.*, Vol. 46, pp. 2775-2780, ISSN 0009-2509

Klamt, F. & Eckert, F. (2000). COSMO-RS: a novel and efficient method for the a priori prediction of thermophysical data of liquids, *Fluid Phase Equilib.* Vol.172, pp. 43–72, ISSN 0378-3812

Konstantinova, N.M.; Motornova, M.S.; Mamontov M.N.; Shishin D.I. & Uspenskaya, I.A. (2011). Partial and integral thermodynamic properties in the sodium chloride-water-1-butanol(*iso*-butanol) ternary systems. *Fluid Phase Equilibria*, in print, ISSN 0378-3812

Liddell, K. (2005). Thermodynamic models for liquid-liquid extraction of electrolytes. *Hydrometallurgy*, Vol. 76, pp.181-192, ISSN 0304-386X

Lide, D.R. (2007-2008) CRC Handbook of Chemistry and Physics, 88th edition, USA, ISBN 0849304881

Lin, S.-T. & Sandler, S.I. (2002). A priori phase equilibrium prediction from a segment contribution solvation model, *Ind. Eng. Chem. Res.* Vol.41, pp. 899–913, ISSN 0888-5885

Lopes, A.; Farelo, F & Ferra, M.I.A. (2001). Activity coefficients of sodium chloride in water-ethanol mixtures: a comparative study of Pitzer and Pitzer-Simonson models. *Journal of Solution Chemistry*, Vol. 30, No. 9, pp. 757-770, ISSN 0095-9782

Lu, X. H. & Maurer, G. (1993). Model for describing activity coefficients in mixed electrolyte aqueous solutions. *AIChE Journal*, Vol. 39, pp. 1527–1538, ISSN 1547-5905

Macedo, E.A., Skovborg, P. & Rasmussen, P. (1990). Calculation of phase equilibria for solutions of strong electrolytes in solvent– water mixtures. *Chem. Eng. Sci.* Vol. 45, pp.875– 882, ISSN 0009-2509

Mamontov, M.N.; Konstantinova; Veryaeva, E.S. & Uspenskaya, I.A. (2010). The thermodynamic properties of solutions of sodium chloride, water, and 1-propanol. *Russian Journal of Physical Chemistry A*, Vol. 84, No. 7, pp. 1098-1105, ISSN 0044-4537

Mazzarese, J. & Popovych, O. (1983). Standard potentials of Li, Na, and K electrodes and transfer free energies of LiCl, NaCl, and KCl in selected ethanol-water and methanol-water solvents. *J. Electrochem. Soc.*, Vol. 130, No. 10, pp. 2032-2037, ISSN 0013-4651

Mock, L.B.; Evans, L. & Chen, C.C. (1990). Thermodynamic representation of phase equilibria of mixed-solvent electrolyte systems. *AIChE Journal*, Vol. 32, pp. 1655-1664, ISSN 1547-5905

Nakamura, T. (2009). Development and Application of Ion-Selective Electrodes in Nonaqueous Solutions. *Analytical sciences*. Vol. 25, pp. 33-40

Omrani, A.; Rostami, A.A. & Mokhtary, M. (2010). Densities and volumetric properties of 1,4-dioxane with ethanol, 3-methyl-1-butanol, 3-amino-1-propanol and 2-propanol binary mixtures at various temperatures. *Journal of Molecular Liquids*, Vol. 157, pp. 18-24, ISSN 0167-7322

Pitzer, K.S. & Mayorga, G. (1973). Thermodynamics of electrolytes. II. Activity and osmotic coefficients for strong electrolytes with one or both ions univalent. *Journal of Physical Chemistry*, Vol. 77, No. 19, pp. 2300-2308, ISSN 1520-6106

Pitzer, K. & Simonson, J.M. (1986). Thermodynamics of multicomponent, miscible, ionic systems: theory and equations. *Journal of Physical Chemistry*, Vol. 90, No. 13, pp. 3005-3009, ISSN 1520-6106

Pol, A. & Gaba, R. (2008). Densities. excess molar volumes, speeds of sound and isothermal compressibilities for {2-(2-hexyloxyethoxy)ethanol+*n*-alkanol] systems at temperatures between (288.15 and 308.15) K. *J. Chem. Thermodynamics*, Vol. 40, pp. 750-758, ISSN 0021-9614

Pretsch, E. (2002). The new wave of ion-selective electrodes. *Analytical Chemistry*, Vol. 74, No. 15, pp. 420A–426A, ISSN 0003-2700

Pungor, E.; Toth, K.; Klatsmanyi, P. G. & K. Izutsu (1983). Application of ion-selective electrodes in nonaqueous and mixed solvents. *Pure&Appl. Chem.*,Vol. 55, pp. 2029-2067, ISSN 0033-4545

Pungor, E. (1998). The Theory of Ion-Selective Electrodes. *Analytical Sciences*. Vol.14, p.249-256

Rashin, A.A. & Honig, B. (1985). Reevaluation of the Born model of ion hydration. *Journal of Physical Chemistry*, Vol. 89, No. 26, pp. 5588-5593, ISSN 1520-6106

Sander, B.; Fredenslund, A. & Rasmussen, P. (1986). Calculation of vapour-liquid equilibria in mixed solvent/salt systems using an extended UNIQUAC equation. *Chem. Eng. Sci.*, Vol. 41, pp. 1171-1183, ISSN 0009-2509

Shishin, D. ; Voskov, A.L. & Uspenskaya, I.A. (2010). Phase equilibria in water-propanol(-1, -2) systems. *Russian Journal of Physical Chemistry A*, Vol. 84, No. 10, pp. 1826-1834, ISSN 0044-4537

Silvester, L.F. & Pitzer, K.S. (1977). Thermodynamics of electrolytes. 8. High-temperature properties, including enthalpy and heat capacity, with application to sodium chloride. *Journal of Physical Chemistry*, Vol. 81, No. 19, pp. 1822-1828, ISSN 1520-6106

Smirnova, N.A. (2003). Phase equilibria modelling in aqueous-organic electrolyte systems with regard to chemical phenomena. *J.Chem.Thermod.*, Vol.35, No.5, pp.747-762, ISSN 0021-9614

Taboada, M. E.; Graber, T. A.; & Cisternas, L. A. (2004). Sodium carbonate extrective crystallization with poly(ethylene glycol) equilibrium data and conceptual process design. *Ind.Eng.Chem.Res.*, Vol. 43, No. 3, pp.835-838, ISSN 0888-5885

Thomsen, K.; Iliuta, M.C. & Rasmussen, P. (2004). Extended UNIQUAC model for correlation and prediction of vapour-liquid-liquid-solid equilibria in aqueous salt systems containing non-electrolytes. Part B. Alcohol (ethanol, propanols, butanols)-water-salt systems. *Chem. Eng. Sci.*, Vol. 59, pp. 3631- 3647, ISSN 0009-2509

Truesdell, A.H. (1968). Activity coefficients of aqueous sodium chloride from 15° to 50°C measured with a glass electrode. *Science*, Vol. 161, No. 3844, pp. 884-886, ISSN 0036-8075

Umezawa, Y.; Buehlmann, P.; Umezawa, K.; Tohda, K. & Amemiya S. (2000). Potentiometric selectivity coefficients of ion-selective electrodes. Part II. Inorganic cations. (IUPAC Technical Report). *Pure Appl. Chem.*, Vol. 72, No. 10, pp. 1851–2082, ISSN 0033-4545

Umezawa, Y.; Umezawa, K.; Buehlmann, P.; Hamada, N.; Aoki, H.; Nakanishi, J.; Sato, M.; Xiao, K.P. & Nishimura Y. (2002). Potentiometric selectivity coefficients of ion-selective electrodes. Part II. Inorganic anions. (IUPAC Technical Report). *Pure Appl. Chem.*, Vol.74, No 6, pp. 923-994, ISSN 0033-4545

van Bochove, G. H.; Krooshof, G.J.P. & de Loos, T.W. (2000). Modelling of liquid-liquid equilibria of mixed solvent electrolyte systems using the extended electrolyte NRTL. *Fluid Phase Equilibria*, Vol. 171, pp. 45-58, ISSN 0378-3812

Veryaeva, E.S.; Konstantinova; N.M., Mamontov, M.N. & Uspenskaya, I.A. (2009). Thermodynamic properties of aqueous-alcoholic solutions of sodium chloride. H_2O-2-C_3H_7OH-NaCl. *Russian Journal of Physical Chemistry A*, Vol. 84, No. 11, pp. 1877-1885, ISSN 0044-4537

Voronin, G. F. (1992). *Thermodynamics and Material Science of Semiconductors. Ch. 31*, Metallurgiya, Moscow, Russia

Wang, P.; Anderko, A. & Young, R.D. (2002). A speciation-based model for mixed-solvent electrolyte systems. *Fluid Phase Equilibria*, Vol. 203, pp. 141-176, ISSN 0378-3812

Wang, P.; Anderko, A.; Springer, R.D. & Young, R.D. (2006). Modeling phase equilibria and speciation in mixed-solvent electrolyte systems: II. Liquid–liquid equilibria and properties of associating electrolyte solutions. *Journal of Molecular Liquids*, Vol. 125, pp. 37 – 44, ISSN 0167-7322

Wikipedia, the free encyclopedia. (22 July 2011). Available from http://en.wikipedia.org/wiki/Ion_selective_electrode

Wroblewski, W. (2005). Ion-selective electrodes. Available from http://csrg.ch.pw.edu.pl/tutorials/ise/

Wu, Y.T.; Zhu, Z.Q.; Lin, D.Q. & Mei, L.H. (1996). Prediction of liquid-liquid equilibria of polymer-salt aqueous two-phase systems by a modified Pitzer's virial equation. *Fluid Phase Equilibria*, Vol. 124, pp. 67-79, ISSN 0378-3812

Zerres, H. & Prausnitz, J.M. (1994). Thermodynamics of phase equilibria in aqueous-organic systems with salt. *AIChE Journal*, Vol. 40 pp. 676-691, ISSN 1547-5905

Zhigang, T.; Rongqi, Z.; & Zhanting, D. (2001). Separation of isopropanol from aqueous solutions by salting-out extraction. *J. Chem. Technol. Biotechnol.* Vol. 76, No. 7, pp. 757-763, ISSN1097-4660

6

Application of Electromotive Force Measurement in Nuclear Systems Using Lead Alloys

Yuji Kurata
Japan Atomic Energy Agency
Japan

1. Introduction

Liquid lead and lead-bismuth eutectic(LBE) are promising candidate materials as coolant of fast reactors with improved safety because of their good thermal-physical and chemical properties. Liquid LBE is also a primary candidate material for high-power spallation targets and coolant of accelerator driven systems (ADSs) for transmutation of long-lived radioactive wastes. In order to apply these liquid lead and LBE to nuclear systems, development of systems for controlling oxygen concentration in the liquid mediums is one of important research subjects (Gromov et al., 1999; Shmatko & Rusanov, 2000; OECD/NEA Handbool, 2007). Liquid lead and LBE are corrosive for steels at high temperatures and likely to cause plugging due to PbO formation in low-temperature components of systems. It is generally accepted that maintaining a certain level of oxygen concentration in them is crucial from the viewpoints of mitigating corrosion attack at high-temperature parts and avoiding PbO formation at low-temperature parts. The active oxygen control within the range between Fe_3O_4 and PbO formation has been proposed (Shmatko & Rusanov, 2000; Li, 2002; OECD/NEA Handbool, 2007).
Effective control of the oxygen concentration requires devices for removing and adding oxygen in liquid lead alloys and oxygen sensors for monitoring the oxygen concentration. It is essential to measure oxygen concentration correctly in lead alloys online for the active oxygen control. Electromotive force measurement using oxygen sensors with a solid electrolyte is a useful means to measure oxygen concentration in liquid lead alloys. Oxygen sensors employing yttria-stabilized zirconia (YSZ) and magnecia-stabilized zirconia (MSZ) as a solid electrolyte enable us to measure oxygen concentration in lead alloys. While oxygen sensors using YSZ as a solid electrolyte and Pt/gas as a reference system are often used in automobile industries, the operating temperature of the sensors is generally rather high. It has been reported that oxygen sensors were used in liquid lead and LBE in Russia where research on nuclear systems using them has been conducted for a long term (Gromov et al., 1999; Shmatko & Rusanov, 2000). The special sensor using YSZ as a solid electrolyte and Mo/Bi-Bi_2O_3 as a reference system gave accurate measurement of oxygen concentration and long service lifetime(Shmatko & Rusanov, 2000). Recently, oxygen sensors for use in liquid lead alloys have been manufactured and examined worldwide.

Konys et al. (Konys et al., 2001, 2004) and Schroer et al. (Schroer et al., 2011) showed that YSZ sensors using a Pt/air reference electrode and a Mo/Bi-Bi_2O_3 reference electrode were promising in LBE as a result of tests of sensors using Pt/air, Mo/Bi-Bi_2O_3 and Mo/In-In_2O_3 reference electrodes. Furthermore, it was reported that the Pt/air reference electrode sensor exhibited better reliability and longer lifetime than the Mo/Bi-Bi_2O_3 reference electrode (Konys et al., 2004). Courouau et al. (Courouau et al., 2002a, 2002b; Courouau, 2004) mainly used the YSZ sensor with the Mo/Bi-Bi_2O_3 reference electrode as a result of tests of several Mo/metal-metal oxide electrodes. Calibration was also conducted for the Mo/Bi-Bi_2O_3 reference electrode sensors manufactured at the laboratory scale (Courouau et al., 2002b). The calibration method firstly proposed by Konys et al. (Konys et al., 2001) is based on measurement of electromotive force (EMF) of the oxygen sensor following temperature variation under the condition close to oxygen saturation in LBE. Calibration methods using Co/CoO or Fe/Fe-oxide equilibrium in liquid LBE besides Pb/PbO equilibrium (oxygen-saturated condition) were also attempted (Schroer et al., 2011). The YSZ sensors with the Mo/Bi-Bi_2O_3 reference electrode have been manufactured and used by other researchers (Li, 2002; Kondo et al., 2006; Num et al., 2008). Although sensors with the Mo/In-In_2O_3 reference electrode have been also developed, the temperature range where measured EMF values agree with theoretical values in the calibration test is very narrow (Konys et al., 2001; Courouau, 2004; Fernandez et al., 2002; Colominas et al., 2004). Furthermore, experience has been reported on use of Russian sensor with the Bi-Bi_2O_3 reference and sensors with the In-In_2O_3 reference in liquid Pb-Bi or liquid Pb (Foletti et al., 2008).

The following two problems have been pointed out in case of using the oxygen sensor with the Mo/metal-metal oxide reference electrode: this type of sensor often exhibits time drift that EMF values change with increase in service time (OECD/NEA Handbool, 2007) and measured EMF values disagree with the theoretical ones (Konys et al., 2001; Colominas et al., 2004). We have also tested YSZ sensors using Pt/air reference and Mo/Bi-Bi_2O_3 reference electrodes in LBE. We had experience that even the sensor with the Pt/air reference electrode exhibited incorrect EMF values in LBE after long-term use as a reliable oxygen sensor. Investigation of the cause of the incorrect outputs and re-activation of the Pt/air reference sensor was reported (Kurata et al., 2010). Importance of interface reaction at electrode/medium and electrocatalyst was pointed out in measurement of electromotive force using YSZ oxygen sensors. Furthermore, calibration methods with high reliability and convenience were required for oxygen sensors for use in liquid lead alloys.

In this chapter, the EMF measurement using the YSZ sensor with the Pt/gas reference system is investigated from basic to practical viewpoints together with re-activation treatments of the YSZ sensor. In addition, the usage record of the YSZ sensors with the Pt/air reference system for a long time and situation of oxygen control in liquid LBE in a static corrosion apparatus are described.

2. Theory of oxygen concentration measurement

Figure 1 shows a principle diagram for measurement of oxygen concentration using a solid electrolyte. A solid electrolyte separates two domains characterized by different oxygen partial pressures. When the electrolyte is an oxygen-ion (O^{2-}) conductor, an electrochemical galvanic cell using a solid electrolyte is presented as follows:

$$Po_2\text{(reference)}//\text{solid electrolyte}//Po_2$$

where P_{O_2}(reference) is the oxygen partial pressure at the reference electrode and P_{O_2} is the oxygen partial pressure at the working electrode. The EMF is formed across the solid electrolyte between the different oxygen partial pressures. The EMF, *E* is expressed as follows according to the Nernst equation:

$$E = \frac{RT}{4F} \ln \frac{P_{O_2}(\text{reference})}{P_{O_2}} \quad (1)$$

where *R* is the gas constant, *T* temperature and *F* the Faraday constant. When the gas containing a given oxygen concentration is used at the reference electrode side, the oxygen partial pressure at the working electrode, P_{O_2} can be calculated using the Eq. (1). The EMF can be measured using a voltmeter shown in Fig.1. A voltmeter with high impedence is recommended in measurement using solid electrolyte sensors.

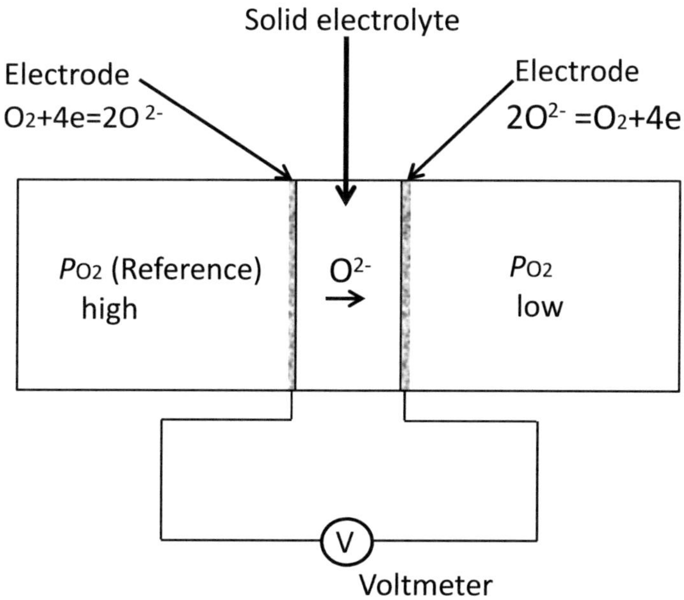

Fig. 1. Principle diagram for measurement of oxygen concentration using a solid electrolyte.

Figure 2 shows a schematic diagram of an oxygen sensor and interface reaction in measurement of oxygen concentration in liquid LBE using YSZ as a solid electrolyte and Pt/gas reference electrode. The following interface reaction occurs on porous Pt as electrocatalyst at the YSZ surface in the reference electrode side:

$$O_2 + 4e = 2O^{2-} \quad (2)$$

Type 304SS was used as a working electrode in liquid LBE. Since oxygen dissolves into liquid LBE as O atom, the following interface reaction occurs at the YSZ surface in the liquid LBE side:

$$2O^{2-} = 2[O] + 4e \quad (3)$$

The oxygen activity, a_o in equilibrium with an oxygen pressure Po_2 is written assuming that dissolution of oxygen into liquid LBE obeys the Henry's law:

$$a_o = \gamma_o C_o = \frac{C_o}{C_o^s} = \left(\frac{P_{o_2}}{P_{o_2}^s} \right)^{\frac{1}{2}} \tag{4}$$

where γ_o is an activity coefficient, C_o the oxygen concentration in LBE, C_o^s the saturated oxygen concentration in LBE and $P_{o_2}^s$ the oxygen concentration in gas in equilibrium with oxygen-saturated LBE. The activity a_o becomes unity when the oxygen dissolved in LBE attains the level of saturation ($C_o = C_o^s$). The saturated oxygen concentration in LBE is calculated using the following Orlov's equation (Gromov et al., 1999):

$$\log C_o^s(wt\%) = 1.2 - \frac{3400}{T} \tag{5}$$

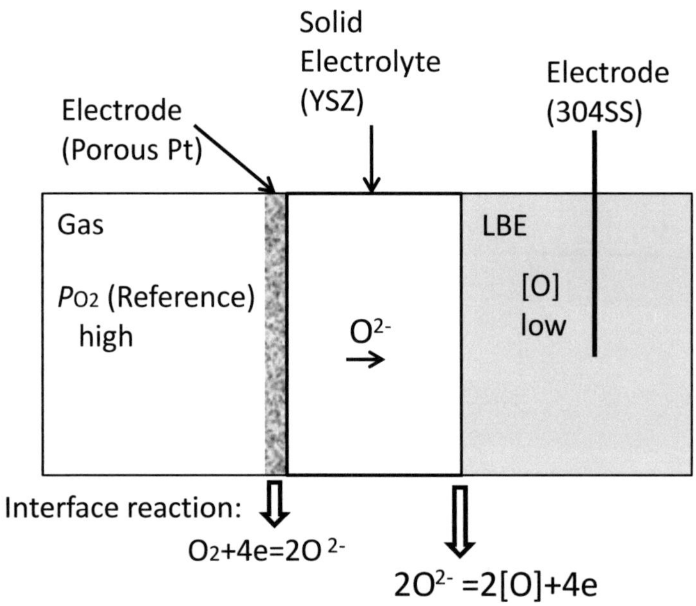

Fig. 2. Reaction at interfaces of solid electrolyte of an oxygen sensor.

Figure 3 shows a schematic diagram to measure oxygen concentration in liquid LBE using two types of oxygen sensors. Oxygen sensors using YSZ as a solid electrolyte and Pt/gas (a) or Mo/Bi-Bi_2O_3 (b) as a reference system were used in this study. When measurement was conducted in LBE, air was used as the reference gas in Pt/gas reference system. The 304SS rod was used as an electrode immersed in LBE. Therefore, the system for measurement in LBE is represented by Pt/air//YSZ//LBE/304SS or Mo/Bi-Bi_2O_3//YSZ//LBE/304SS. The relationship between the EMF and the oxygen concentration in LBE has been calculated for

these two reference electrode sensors using standard Gibbs energy of PbO and Bi_2O_3 (Courouau et al., 2002a; Konys et al., 2004). The equation derived by Courouau et al. (Courouau et al., 2002a) was used in this study.
For Pt/air reference sensor

$$E_{\mathrm{Saturation}}=1.129-5.858\mathrm{x}10^{-4}T \tag{6}$$

$$E=0.791-4.668\mathrm{x}10^{-4}T-4.309\mathrm{x}10^{-5}T\ln C_{\mathrm{O}} \tag{7}$$

For Mo/Bi-Bi_2O_3 reference sensor

$$E_{\mathrm{Saturation}}=0.128-6.368\ \mathrm{x}10^{-5}T \tag{8}$$

$$E=-0.210+5.538\mathrm{x}10^{-5}T-4.309\mathrm{x}10^{-5}T\ln C_{\mathrm{O}} \tag{9}$$

Thermoelectric voltages occur between Mo wire and austenitic stainless steels such as 304SS in measurement using the Mo/Bi-Bi_2O_3 reference electrode sensor. The influence of the thermoelectric valgtages on measurement was investegated in detail (Schroer et al., 2011).

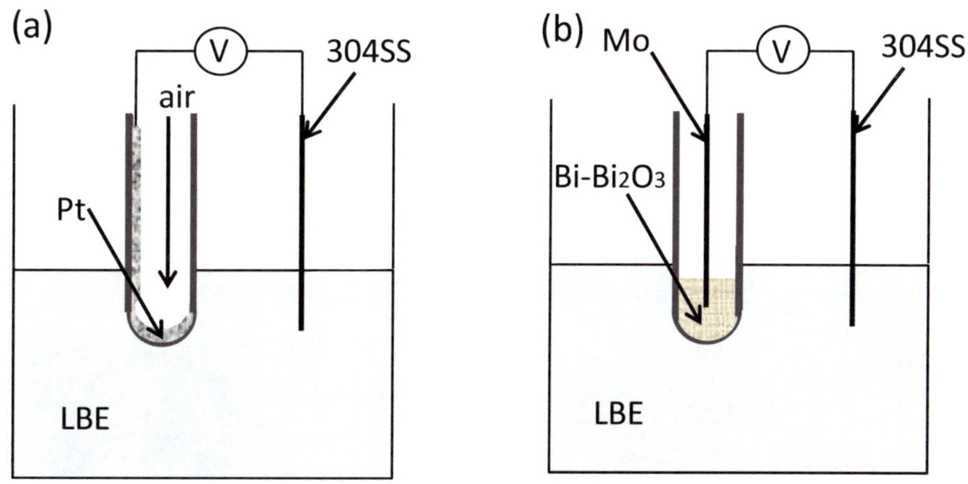

Fig. 3. Schematic diagram of oxygen sensors in liquid LBE: (a) Pt/air reference system and (b) Mo/Bi-Bi_2O_3 reference system.

3. Electromotive force measurement using oxygen sensors

While YSZ sensors with the Pt/gas reference electrode and the Mo/Bi-Bi_2O_3 reference electrode were prepared, the former was mainly used in this study. The Pt/gas reference sensor made by Sukegawa Electric Co., Ltd. was the one-end closed YSZ tube with outer diameter of 15mm and inner diameter of 11mm. The Pt/gas reference system was put inside the YSZ tube. The inner Pt electrode was made by a process of painting Pt-paste inside the YSZ tube and baking it. The porous Pt electrode made through this process has good catalytic activity that enables to measure oxygen concentration at lower temperatures. The fully stabilized zirconia (ZrO_2) with 8 mol% Y_2O_3 produced by Nikkato Corp. was used as a solid electrolyte on account of its good electronic behavior and thermo-mechanical performance. The Mo/Bi-Bi_2O_3 reference sensor was a sensor using the one-end closed tube of YSZ with the sizes of 8mm in outer diameter, 5mm in inner diameter and 300mm in length. The Mo/Bi-Bi_2O_3 reference electrode was made inside the YSZ tube in our laboratory. The ratio of Bi to Bi_2O_3 was 9:1 in weight. The upper part of the YSZ tube was sealed using alumina cement. This sensor with the Mo/Bi-Bi_2O_3 reference electrode was similar to the Mo/Bi-Bi_2O_3 reference sensor manufactured in other institutes (Courouau et al., 2002b; Konys et al., 2004; Kondo et al., 2006).

The following two methods were employed for calibration of oxygen sensors: (1) comparison between measured EMF values and theoretical ones using two kinds of gases with different oxygen concentrations for the reference electrode and the working electrode, and (2) comparison between measured EMF values and theoretical ones in LBE with the parameter of temperature under the condition close to oxygen saturation in LBE. The advantage of the former method is easiness of preparing the reliable working electrode with the correct oxygen concentration in gas in case of employing a ceramic vessel. The latter method has been often employed as a calibration test in LBE (Konys et al., 2001; Courouau et al., 2002b). An electrometer with high impedence of $10^{14}\Omega$ was used for measurement of the EMF both in gas and in LBE. Schroer et al. conducted calibration tests using not only Pb/Pb-monoxiside (PbO) but also Co/Co-monoxide (CoO) and Fe/Fe-oxide equilibria in liquid LBE (Schroer et al., 2011).

3.1 Measurement of oxygen concentration in gas

Figure 4 shows a schematic drawing (a) and appearance (b) of the Pt/gas reference sensor used in measurement of oxygen concentration in gas (Kurata et al., 2010). Platinum paste was painted on the lower part of the outer YSZ surface to measure oxygen concentration in gas. In this test, 10.45%O_2-He gas was used as reference gas and 502ppmO_2-He gas as working gas. The temperature range was 350°C - 600°C and the temperature was kept for about 24h to investigate change of the EMF values at each temperature.

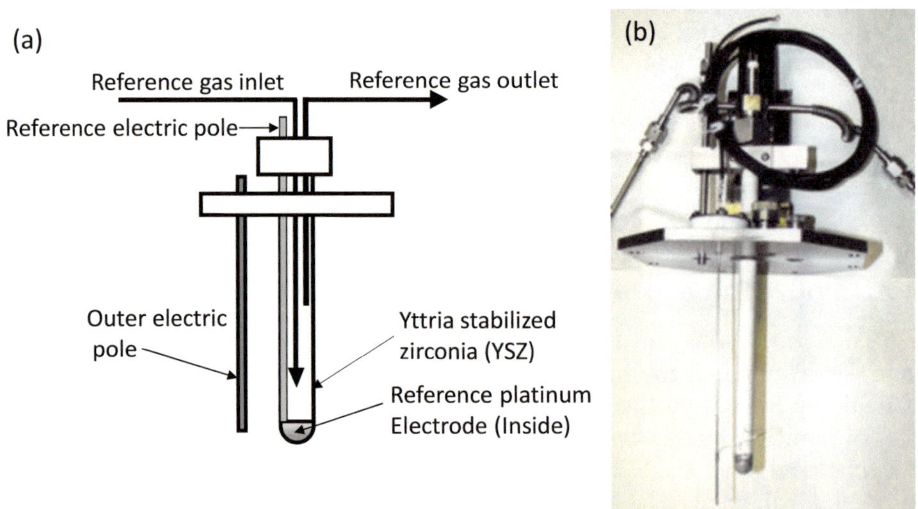

Fig. 4 Schematic drawing (a) of the YSZ oxygen sensor with Pt/gas reference system and appearance(b) of the YSZ oxygen sensor with outer Pt electrode for measurement of oxygen concentration in gas (Kurata et al., 2010).

The relationship between the EMF and tempeerature is shown in Fig. 5 (Kurata et al., 2010). The theoretical line calculated from Eq. (1) is also drawn in this figure. The EMF values approach the theoretical line of the Nernst relation while it seems to take time to attain the stable outputs below 500°C. This calibration method in gas was often used to investigate correctness of Pt/gas reference sensors.

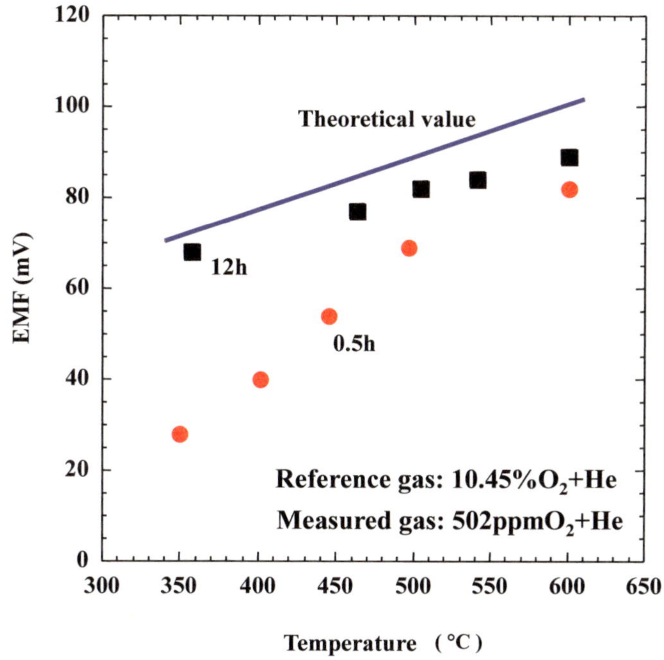

Fig. 5. Relationship between EMF and temperature measured in gas using Pt/gas reference sensor (Kurata et al., 2010).

3.2 Estimation of outputs of the Pt/gas reference sensor in liquid LBE

An apparatus for corrosion tests in LBE (Kurata et al., 2008) was used for the calibration test in LBE. Components contacting liquid LBE were made of quartz. About 7kg of LBE was put into the pot and melted under Ar cover gas with purity of 99.9999% for the calibration test in LBE. The chemical compositions of LBE were 55.60Bi-0.0009Sb-0.0002Cu-0.0001Zn-0.0005Fe-0.0007As-0.0005Cd-0.0001Sn-Bal.Pb(wt%). Initial oxygen content in LBE was usually from 10^{-4} to 10^{-3} wt% in this treatment. Figure 6 depicts a photo showing the Pt/gas reference oxygen sensor under measurement in liquid LBE. A thin PbO film was observed on the surface of the liquid LBE with pure Ar cover gas at 450°C.

Figure 7 shows the relationship between EMF and temperature measured in LBE using Pt/gas reference sensor (Kurata et al., 2010). Air was used as reference gas of the YSZ oxygen sensor. Open circles indicate EMF values measured in oxygen-saturated LBE with pure Ar cover gas. The theoretical line calculated from Eq. (6) for the oxygen-saturated LBE is written with a thick solid line. The measured EMF values are almost on the theoretical line for the oxygen-saturated LBE above 450°C. From the measured EMF value at 550°C, it is estimated that the oxygen concentration in the LBE is about 10^{-3}wt%. The measured EMF

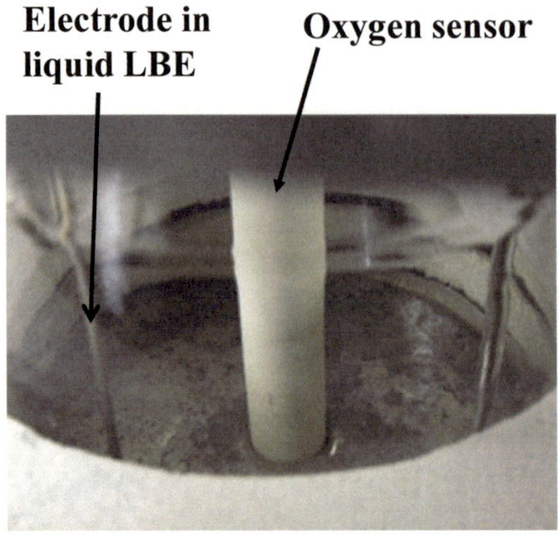

Fig. 6. Photo showing the oxygen sensor under measurement in liquid LBE.

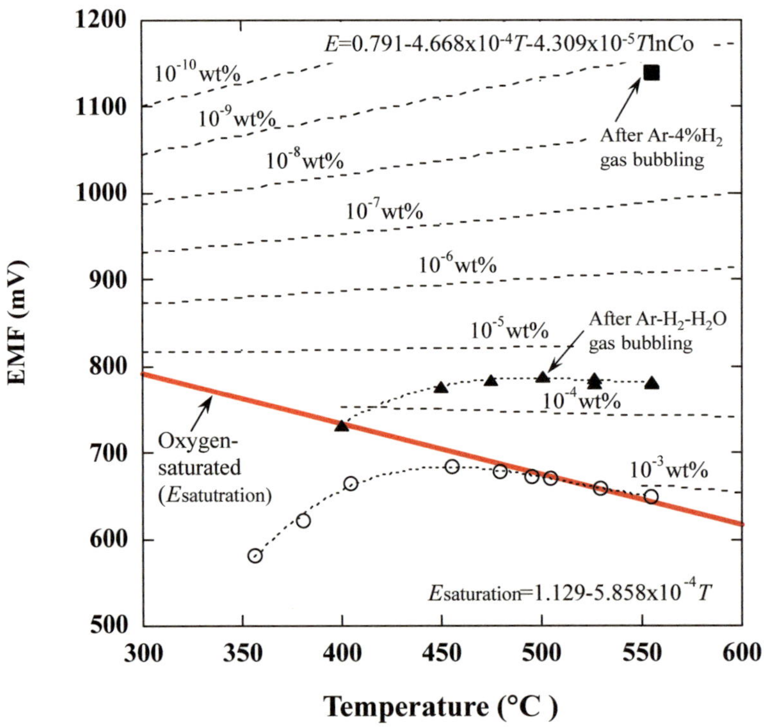

Fig. 7. Relationship between EMF and temperature measured in LBE using Pt/gas reference sensor (Kurata et al., 2010).

values are much lower than the theoretical line below 400°C. Furthermore, the measured EMF value attained the stable one in LBE above 450°C in short time. From the calibration test using Pb/PbO equilibrium in liquid LBE, it is possible to use the Pt/air reference sensor above 450°C in liquid LBE. Solid triangles indicate EMF values measured in LBE after Ar-H_2-H_2O gas bubbling. These data were obtained in oxygen-unsaturated LBE. The theoretical lines calculated from Eq. (7) are drawn for the EMF values of oxygen concentrations of 10^{-3}wt% to 10^{-10}wt% in LBE. Regarding EMF values measured in LBE after Ar-H_2-H_2O gas bubbling, the slope and the magnitude above 450°C are identical with the expected values for LBE with dissolved oxygen concentration of about 3x10^{-5}wt%. A solid square shows the EMF value measured in LBE after Ar-4%H_2 gas bubbling. Oxygen concentration of 10^{-9}wt% in LBE can be measured using the Pt/gas reference sensor. Konys et al. and Schroer et al. also showed validation of oxygen sensors from calibration tests in saturated and unsaturated LBE (Konys et al., 2001; Schroer et al., 2011). On the basis of the results obtained in the present test, it is found that the Pt/air reference sensor enables us to measure oxygen concentration correctly in LBE above 450°C. The appearance of the oxygen sensor after the test in LBE is shown in Fig. 8 (Kurata et al., 2010). Since much LBE adheres to the YSZ surface of the sensor, it is clear that the YSZ surface is wet well with liquid LBE.

Fig. 8. Appearance of the Pt/gas reference sensor after test in liquid LBE (Kurata et al., 2010).

3.3 Re-activation of oxygen sensor

The Pt/air reference sensor, which exhibited good performance, had been used in LBE for about 6500h. A comparison test was conducted in LBE for the Pt/air reference sensor after long-term use and the Mo/Bi-Bi_2O_3 reference sensor produced in our laboratory. An apparatus shown in Fig. 9 was used for the comparison test of oxygen sensors. The vessel of the apparatus was made of 304SS and outputs from three sensors in liquid LBE could be compared. About 60 kg of LBE was used in the comparison test using 304SS vessel. The procedure similar to that in section 3.2 was employed in the comparison test in liquid LBE. Figure 10 shows the relationship between the EMF and temperature measured in LBE using Pt/air reference and Mo/Bi-Bi_2O_3 reference sensors (Kurata et al., 2010). In the same way as Fig. 7, the theoretical lines calculated from Eq. (6) for the Pt/air reference sensor and from Eq. (8) for the Mo/Bi-Bi_2O_3 reference sensor in the oxygen-saturated LBE are drawn in this figure. It is a surprise that the measured EMF values in LBE with Ar cover gas are much higher than each theoretical line of oxygen-saturated LBE because oxygen concentration in calibration tests using LBE with Ar cover gas has been constantly in a range of 10^{-4} to 10^{-3} wt%. Therefore, it is necessary to examine whether EMF values measured by both sensors in this test showed correct oxygen concentration or not. Since Ar cover gas does not contain a reducing gas component, it is considered that fresh LBE used in the test contained oxygen of 10^{-4} to 10^{-3} wt%. In the case of the Pt/air reference sensor, the slope of the relationship between the EMF value and temperature is similar to that of the theoretical line above 400°C. Similar trend is also observed in the case of the Mo/Bi-Bi_2O_3 reference sensor above 350°C. These results suggest that oxygen concentration in LBE used in the test was close to saturated oxygen concentration.

Fig. 9. Photo of the comparison apparatus of oxygen sensors. Three sensors can be compared in liquid LBE.

If LBE is oxygen-saturated, it is considered that both sensors exhibited high EMF outputs including somewhat bias voltage. Courouau et al. showed time drift of Mo/metal-metal oxide electrode sensors and presented several hypotheses to explain the cause of the time drift: alteration of the interface of the electrode(working or reference) by oxide deposition, reaction with LBE or the liquid metal reference, or alteration of YSZ affecting eventually the electrode potential (Courouau, 2004; OECD/NEA Handbool, 2007). When the magnitude of the effect on the electrode potential is constant, the alteration can produce constant bias voltage. The comparison tests using the same quartz pot as that in section 3.2 were repeated for the Pt/air reference and Mo/Bi-Bi_2O_3 reference sensors. According to some analyses of results (Kurata et al., 2010), the bias voltage was not always constant although values of bias voltage varied from 200mV to 260mV in repeated calibration tests. Therefore, it is generally difficult to employ the correction method by the constant bias voltage.
Figure 11 depicts appearance of oxygen sensors after measurement in liquid LBE using the comparison apparatus made of 304SS. The black soot of Pb and Bi deposited with LBE on the YSZ surface. There were various surface conditions on the YSZ of the sensors after the comparison tests in LBE. While the YSZ surface was often wet, it was not wet sometimes in particular at low temperatures. Both sensors with Pt/air reference and Mo/Bi-Bi_2O_3 reference electrodes exhibited higher EMF values above 200mV than the theoretical ones above 400°C in all cases after the comparison test.

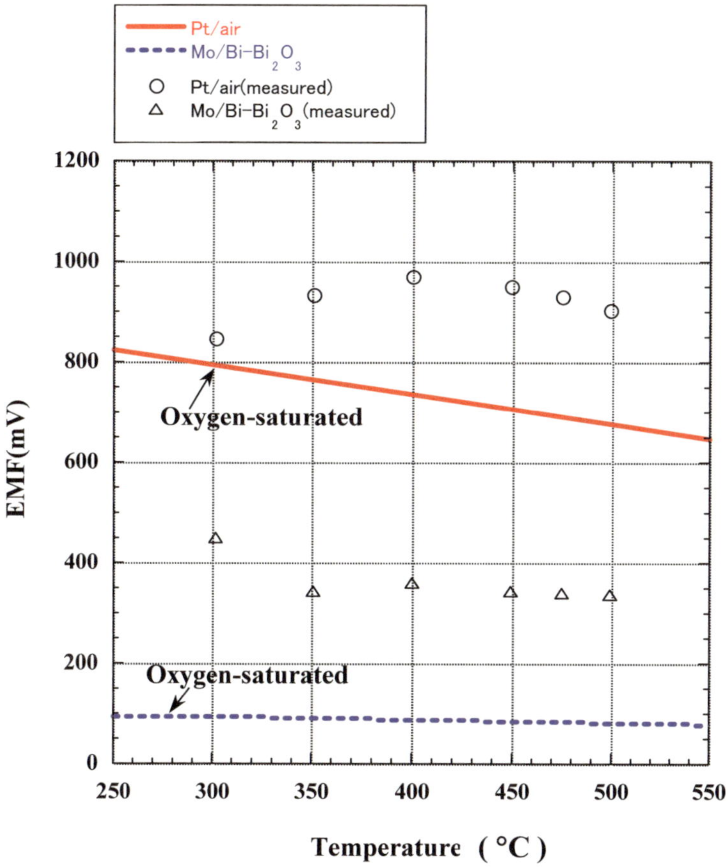

Fig. 10. Relationship between EMF and temperature measured in LBE using Pt/air reference and Mo/Bi-Bi_2O_3 reference sensors (Kurata et al., 2010).

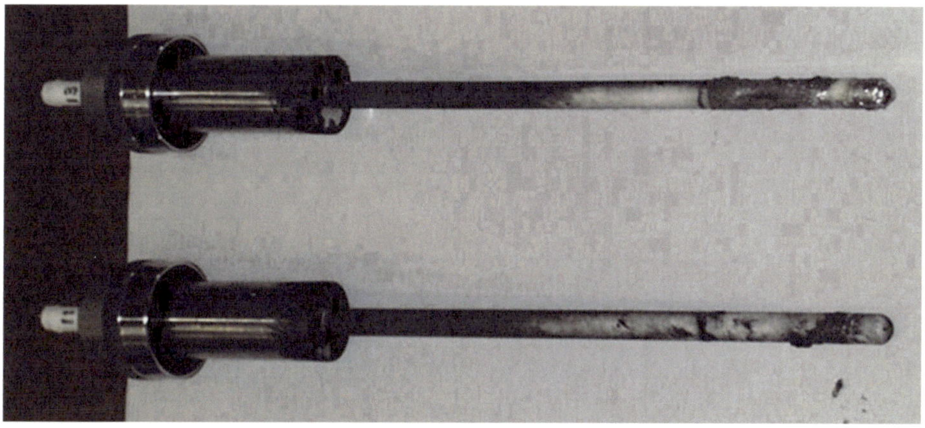

Fig. 11. Appearance of oxygen sensors after measurement in liquid LBE using the comparison apparatus of oxygen sensors.

Investigation of re-activation treatment is one of important research subjets. Some re-activation treatments were attempted for the Pt/gas reference sensor that exhibited incorrect outputs. First of all, the outer surface of the YSZ tube was washed with a nitric acid to remove adherent LBE and black soot. Figure 12 depicts appearance of the Pt/gas reference sensor after cleaning with a nitric acid. Although most of LBE and black soot seem to be removed, there are some black spots left. Figure 13 shows results of the calibration test in liquid LBE using a quartz pot after the cleaning with a nitric acid (Kurata et al., 2010). The Pt/air reference sensor after the washing exhibits higher EMF values by about 220mV than the theoretical line of oxygen-saturated LBE above 450°C. Therefore, it is not capable of recovering the ability of the Pt/gas reference sensor by the method of cleaning with a nitric acid.

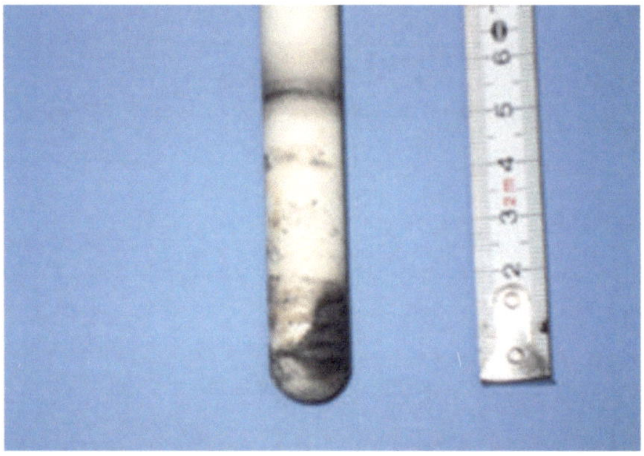

Fig. 12. Photo showing appearance of the Pt/gas reference sensor after cleaning with a nitric acid.

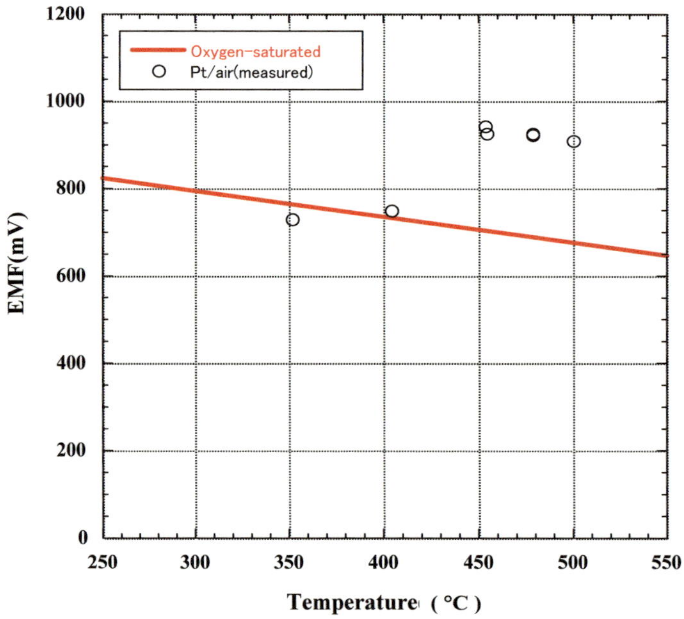

Fig. 13. EMF measurement in LBE using the Pt/gas reference sensor after cleaning with a nitric acid (Kurata et al., 2010).

The Pt-treatment was made on the outer surface of the YSZ next. This treatment is required to measure oxygen concentration in gas and also useful to clean and activate the YSZ surface. Figure 14 shows the relationship between EMF and temperature measured in gas using Pt/gas reference sensors after Pt-treatment (Kurata et al., 2010). The old sensor is the Pt/gas reference one that exhibited inncorrect outputs after the comparison tests. The calibration test in gas was also conducted for a new Pt/gas reference sensor. In this calibration test, air was used as reference gas and 504ppmO_2-He gas as working gas. The old Pt/gas reference sensor that exhibited incorrect outputs in LBE indicates the EMF values almost equal to the theoretical ones calculated from Eq. (1) above about 400°C. Furthermore, a new Pt/gas reference sensor exhibits the EMF values almost equal to the theoretical ones above 450°C. The following three causes are considered for the poor condition of the old Pt/gas reference sensor that exhibited incorrect outputs in LBE: (1)failure or degradation of the inner Pt/gas reference system, (2)alteration of YSZ itself and (3)alteration of the outer YSZ surface in contact with liquid LBE. If the cause of the incorrect outputs is failure or degradation of the inner Pt/gas reference system or alteration of YSZ itself, then the old sensor exhibits incorrect outputs in gas. Since the old sensor exhibits correct outputs in gas, the cause of the incorrect outputs seems to be alteration of the outer YSZ surface.

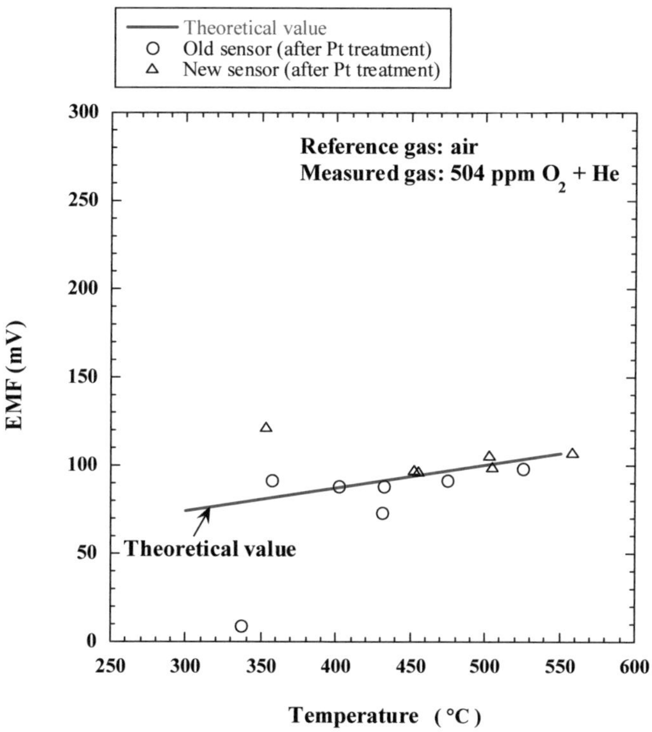

Fig. 14. Relationship between EMF and temperature measured in gas using Pt/gas reference sensors after Pt-treatment (Kurata et al., 2010).

The calibration test in liquid LBE was conducted for both Pt/gas reference sensors after the test in gas. The sensors were soaked into liquid LBE as the state of Pt-treatment on the outer YSZ surface. Results of the calibration test in liquid LBE are shown in Fig.15 (Kurata et al.,

2010). The old sensor after the Pt-treatment exhibits the EMF values almost equal to the theoretical line of oxygen-saturated LBE above 450°C. The new Pt/gas reference sensor after the Pt-treatment also indicates similar behavior. Considering these points into account, the Pt-treatment, which enables us to measure oxygen concentration in gas, seems to play a useful role for measuring oxygen concentration in LBE. As shown in Fig.2, it is essential to continue the interface reaction of Eq. (3) at the outer YSZ surface in contact with liquid LBE in order to measure oxygen concentration correctly using the YSZ electrolyte. The Pt electrode made on the YSZ surface has catalytic characteristics in gas for dissociation reaction of Eq.(2) and enables us to measure oxygen concentration at lower temperatures. The Pt-treatment is composed of painting Pt paste and baking. This treatment produces the porous Pt electrode and clean the YSZ surface. The YSZ interface becomes the activated state due to assistance of the Pt electrode after attaining electrochemical equilibrium under a gas environment. The role of the Pt electrode is a little different in LBE while the Pt electrode is an excellent electrocatalyst in gas. There are two possibilities: improvement of wetting and formation of clean and activated YSZ surface. Dissolution of Pt on YSZ into LBE brings improvement of wetting of the YSZ surface by liquid LBE. Once the YSZ surface is activated in gas by the Pt electrode, the activated state of the YSZ surface will continue in LBE and be useful to promote dissociation reaction at the YSZ surface. The importance of the electrochemical equilibrium of Eq. (3) at the interface between the YSZ and LBE is obvious in measurement of oxygen concentration in liquid LBE using the YSZ electrolyte. When there are highly catalytic electrodes to promote dissociation reaction of Eq. (3) in LBE, it will be possible to conduct measurement of oxygen concentration in LBE with high reliability at low temperatures. Newly developed Ir-C composit and Ru-C composit electrodes, which could operate in a gas environment at temperatures below 300°C due to the excellent catalytic activity (Goto et al., 2001; Sakata et al., 2007), may be useful for the YSZ oxygen sensor with high accuracy and reliability at low temperatures.

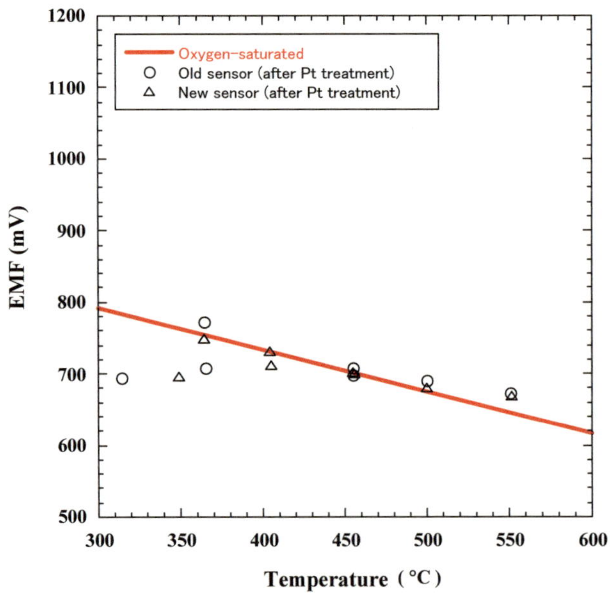

Fig. 15. Relationship between EMF and temperature measured in LBE using Pt/gas reference sensors after Pt-treatment (Kurata et al., 2010).

3.4 Long-term performance of Pt/gas reference sensors

The oxygen sensors with a solid electrolyte of YSZ and a Pt/gas reference electrode have exhibited good performance and been used for a long time in our laboratory to measure oxygen concentration in liquid LBE. Figure 16 shows usage records of the oxygen sensors with the Pt/gas reference electrode. In particular, the sensor-1 was re-activated with Pt-treatment after exhibiting incorrect outputs for some time in liquid LBE. The sensor-1 was used repeatedly for static corrosion tests at 450°C to 550°C through performance tests under gas and liquid LBE environments after re-activation treatment. The cumulative usage time of the sensor-1 attains about 17000h. Controlled oxygen concentration conditions in liquid LBE were 10^{-8} to 10^{-3} wt% and a temperature range in measurement using the oxygen sensor was from 350 to 550°C. In addition, the sensor-2 has been used in static corrosion tests of various steels for 5300h. As mentioned above, the oxygen sensor with the Pt/gas reference electrode can be used as a useful sensor to measure oxygen concentration in liquid LBE.

Figure 17 depicts variation of electromotive force of the oxygen sensor and the oxygen concentration in liquid LBE during the corrosion test at 550°C for 1000h. The oxygen concentration was controlled using Ar-H_2-H_2O gas flow over liquid LBE during the corrosion test. The H_2/H_2O ratio was changed depending on the oxygen concentration in LBE. The range of H_2/H_2O ratio employed in this corrosion test was from 0.2 to 1.0. As shown in this figure, the oxygen concentration in LBE was controlled in the range of 10^{-6} to 10^{-4} wt% for 1000h. The accuracy of the oxygen sensor with the Pt/gas reference electrode was checked before the start of each corrosion test through the process that the output of the oxygen sensor at 450°C or 500°C was compared with the theoretical one of oxygen-saturated LBE. This procedure in LBE is a convenient and reliable calibration method of oxygen sensors. The YSZ sensors with the Pt/air reference electrode also showed good durability and reliability for a long time in LBE loop facilities at 550°C (Konys et al., 2004). In static corrosion tests, the YSZ sensors with the Pt/air reference electrode have been used for a long time in our laboratory as the reliable oxygen sensor in liquid LBE.

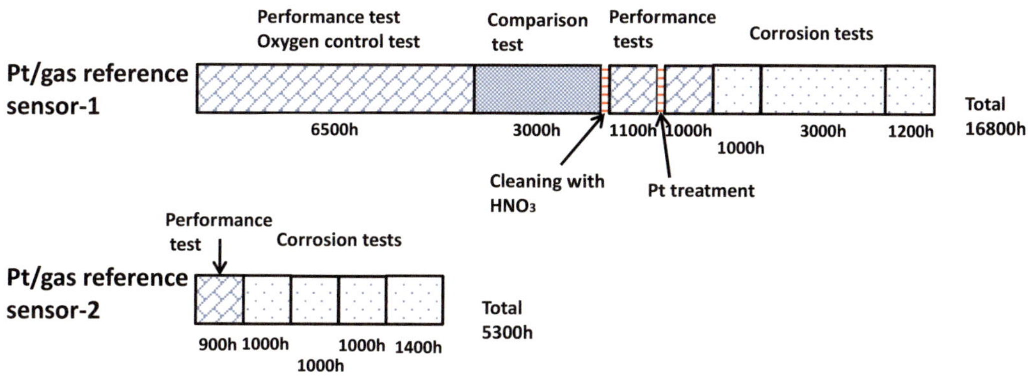

Fig. 16. Usage records of oxygen sensors with Pt/gas reference electrode.

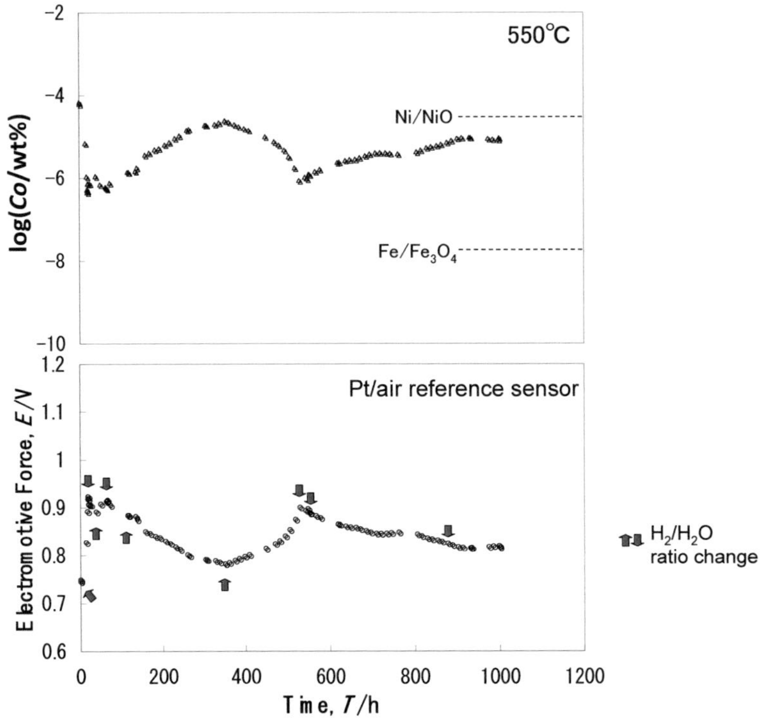

Fig. 17. Oxygen concentration in liquid LBE during the corrosion test at 550°C for 1000h.

4. Concluding remarks

Electromotive force measurement using oxygen sensors with a solid electrolyte of YSZ and a Pt/gas reference electrode is a useful and reliable means to measure oxygen concentration correctly in liquid LBE online. The accuracy of Pt/gas reference sensors was validated in terms of EMF measurements in gas with known oxygen concentration and in oxygen-saturated LBE. The Pt/gas reference sensors can be certainly used to measure oxygen concentration in liquid LBE at least above 450°C. It occurs that even the YSZ sensor with the Pt/gas reference electrode exhibits incorrect outputs on account of contamination such as deposition of black soot etc. on the outer surface of the YSZ. It is possible to re-activate the YSZ sensor, which exhibited incorrect outputs, by means of Pt-treatment on the outer YSZ surface. The attainment of the electrochemical equilibrium at the interface between YSZ and LBE is important in EMF measurement using the YSZ sensor to estimate oxygen concentration in liquid LBE. The YSZ sensors with the Pt/gas reference electrode have been used for a long time as a reliable oxygen sensor to monitor oxygen concentration in liquid LBE online.

5. Acknowledgment

The author would like to thank Drs. M. Futakawa and H. Oigawa at JAEA and Dr. Y. Abe at Sukegawa Electric Co. Ltd. for their encouragement.

6. References

Colominas, S.; Abella, J. & Victori, L. (2004). Characterization of an Oxygen Sensor Based on In/In_2O_3 Reference Electrode. *Journal of Nuclear Materials*, Vol. 335 (November 2004). pp. 260-263, ISSN 0022-3115

Courouau, J.-L. et al. (2002a). Impurities and Oxygen Control in Lead Alloys. *Journal of Nuclear Materials*, Vol.301 (February 2002), pp. 53-59, ISSN 0022-3115

Courouau, J.-L.; Deloffe, P. & Adriano, R. (2002b). Oxygen Control in Lead-bismuth Eutectic: First Validation of Electrochemical Oxygen Sensors in Static Conditions. *Proceedings of Structural Materials for Hybrid Systems: A Challenge in Metallurgy, Journal de Physique IV France*, Vol.12 (September 2002). pp. 141-153, ISBN 2-86883-617-8

Courouau, J.-L. (2004). Electrochemical Oxygen Sensors for On-line Monitoring in Lead-bismuth Alloys: Status of Development. *Journal of Nuclear Materials*, Vol.335 (November 2004), pp. 254-259, ISSN 0022-3115

Fernandez, J.A. et al. (2002). Development of an Oxygen Sensor for Molten 44.5% Lead-55.5% Bismuth Alloy. *Journal of Nuclear Materials*, Vol. 301 (February 2002), pp. 47-52, ISSN 0022-3115

Foletti, C.; Gessi, A. & Benamiti, G. (2008). ENEA Experience in Oxygen Measurements. *Journal of Nuclear Materials*, Vol. 376 (June 2008), pp. 386-391, ISSN 0022-3115

Goto, T.; Ono, T. & Hirai, T. (2001). Electrochemical Properties of Iridium-Carbon Nano Composite Films Prepared by MOCVD. *Scripta Materialia*.Vol. 44 (May 2001), pp. 1187-1190, ISSN 1359-6462

Gromov, B. F. et al. (1999). The Problems of Technology of the Heavy Liquid Metal Coolants (Lead-bismuth and Lead). *Proceedings of the Conference of the Heavy Liquid Metal Coolants in Nuclear Technology, HLMC'98*, October 5-9, 1998, Obninsk, Russia, (1999) pp. 87-100

Kondo, M. et al. (2006). Study on Control of Oxygen Concentration in Lead-bismuth Flow Using Lead Oxide Particles. *Journal of Nuclear Materials*, Vol. 357 (October 2006), pp. 97-104, ISSN 0022-3115

Konys, J. et al. (2001). Development of Oxygen Meters for the Use in Lead-bismuth. *Journal of Nuclear Materials*. Vol. 296 (July 2001) pp. 289-294, ISSN 0022-3115

Konys, J. et al. (2004). Oxygen Measurements in Stagnant Lead-bismuth Eutectic Using Electrochemical Sensors. *Journal of Nuclear Materials*, Vol. 335 (November 2004), pp. 249-253, ISSN 0022-3115

Kurata, Y.; Futakawa, M. & Saito, S. (2008). Corrosion Behavior of Steels in Liquid Lead-bismuth with Low Oxygen Concentrations. *Journal of Nuclear Materials*, Vol. 373 (February 2008), pp. 164-178, ISSN 0022-3115

Kurata, Y. et al. (2010). Characterization and Re-activation of Oxygen Sensors for Use in Liquid Lead-bismuth. *Journal of Nuclear Materials*, Vol. 398 (March 2010), pp. 165-171, ISSN 0022-3115

Li, N. (2002). Active Control of Oxygen in Molten Lead-bismuth Eutectic Systems to Prevent Steel Corrosion and Coolant Contamination. *Journal of Nuclear Materials*, Vol. 300 (January 2002), pp. 73-81, ISSN 0022-3115

Num, H. O. et al. (2008). Dissolved Oxygen Control and Monitoring Implementation in the Liquid Lead-bismuth Eutectic Loop: HELIOS. *Journal of Nuclear Materials*, Vol. 376 (June 2008), pp. 381-385, ISSN 0022-3115

OECD/NEA Handbool. (2007). *Handbook on Lead-bismuth Eutectic Alloy and Lead Properties, Materials Compatibility, Thermal-hydraulics and Technologies* (2007), pp. 151-165, ISBN 978-92-64-99002-9
http://www.nea.fr/html/science/reports/2007/nea6195-handbook.html

Sakata, M.; Kimura, T. & Goto, T. (2007). Microstructures and Electrical Properties of Ru-C Nano-composite Films by PECVD. *Materials Transactions*, Vol. 48 (January 2007), pp. 58-63, ISSN 1345-9678

Schroer, C. et al. (2011). Design and Testing of Electochemical Oxygen Sensors for Service in Liquid Lead Alloys. *Journal of Nuclear Materials*,Vol. 415 (August 2011), pp.338-347, ISSN 0022-3115

Shmatko, B. A. & Rusanov, A. E. (2000). Oxide Protection of Materials in Melts of Lead and Bismuth. *Materials Science*, Vol. 36, No. 5 (May 2000), pp. 689-700, ISSN 1068-820X

7

Electromotive Force Measurements in High-Temperature Systems

Dominika Jendrzejczyk-Handzlik and Krzysztof Fitzner
AGH University of Science and Technology, Laboratory of Physical Chemistry and Electrochemistry, Faculty of Non-Ferrous Metals, Krakow, Poland

1. Introduction

Stability of phases existing in chemical systems is determined by its Gibbs free energy designated as G. The relative position of Gibbs free energy surfaces in the G–T–X (composition) space determines stability ranges of respective phases yielding a map called the phase diagram. Since the knowledge of phase equilibria is essential in designing new materials, determination of Gibbs free energy for respective phases is being continued on both theoretical as well as experimental ways. While in principle chemical potentials of pure substances are needed to derive Gibbs free energy of formation of the stoichiometric phases, it is not the case for the phase (solid or liquid) with variable composition. As an example, in Fig 1, ΔG_m for three different systems is shown. Fig.1a shows free energy of formation of the intermetallic, stoichiometric Mg_2Si phase recalculated per one mole of atoms (Turkdogan, 1980). Fig1b illustrates Gibbs energy of formation of one mole of liquid In-Pb solution (Hultgren, 1973). Finally, Fig.1c demonstrates Gibbs energy of formation of the solid phase, wustite ('FeO') (Spencer & Kubaschewski, 1978).

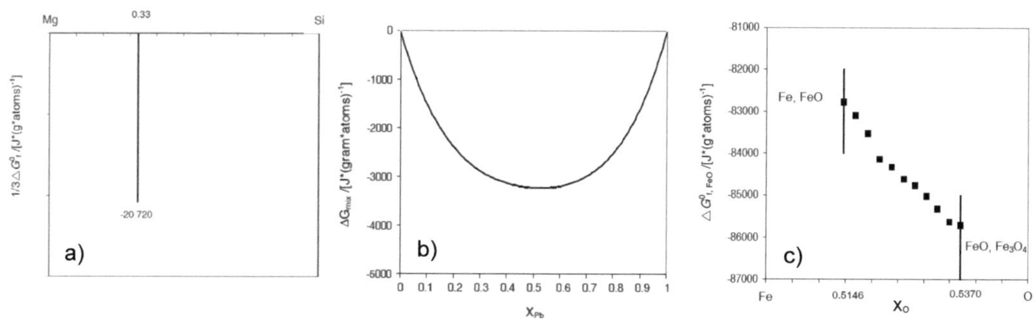

Fig. 1. Gibbs free energy of formation of a) stoichiometric Mg_2O phase, b) liquid In-Pb solution, c) solid phase 'FeO' wustite.

In the last two cases experimental information about chemical potentials of both components in the solution (partial Gibbs energies) was necessary to obtain ΔG_m vs. composition dependencies at fixed temperature.

In general, there are four experimental methods, namely calorimetry, vapour pressure, electrochemical and phase equilibration, one can use to obtain thermodynamic functions, which describe properties of respective phases, solid or liquid. Calorimetry is an indispensable tool to measure enthalpy changes , but its weakness consists in the fact that a number of calorimetric measurements must be combined in order to obtain Gibbs energy changes. Vapour pressure methods (both static and dynamic) cover wide range of vapour pressures from which standard free energy change as well as activities (chemical potentials) of components in the solution can be obtained. The most powerful modification of this technique is effusion method combined with mass spectrometry, which identifies and gives the partial pressure of all species present in the gas phase. Another advantage is the temperature of experiments, which cannot be matched by any other method. Partial Gibbs energy can also be derived from the investigation of equilibrium between different phases. In most cases the success of this method relies heavily on the accuracy of chemical analysis of phases involved in chemical equilibrium. Finally, an electrochemical method (so-called e.m.f. method) which is based on properly designed electrochemical cell can supply information about chemical potential of the components in any phase: gaseous, liquid or solid. Though, undoubtedly calorimetry is the most precise and direct method to measure heat effects of chemical changes i.e. enthalpy changes, chemical potential is needed to derive Gibbs free energy change. Both, e.m.f. and vapour pressure methods can yield chemical potential of the component through its activity measurements, but this approach has one weakness. Usually, we can measure activity for only one component in the solution. In good, old days Gibbs-Duhem equation was used to solve this problem and to derive activities for other components. In the age of modeling and computers this problem is solved much faster and the desired expression:

$$G_m = X_A \mu_A + X_B \mu_B + \ldots \tag{1}$$

(where X_i denotes mole fraction and μ_i denotes chemical potential)
for Gibbs free energy of one mole of the either solid or liquid phase can be easily obtained.
As far as the determination of the partial Gibbs energy of the components is concerned, in our opinion the electrochemical method may be considered as the most accurate one, though not without many traps. Ben Alcock used to say that e.m.f. method is the best method to derive activity of the component in the solution if…….works. This humorous and even a little spiteful comment is perhaps a good reason to discuss the principles and the range of applicability of this method.

2. Principles

When an electronic conductor (metal, semiconductor, polymer) is brought into equilibrium with ionic conductor (liquid electrolyte solution, molten salt, solid electrolyte, etc.) an interface between these two phases is created. Then, due to the charge separation between these two phases, the interface is charged and an electric potential difference Φ across the interface builds up. Such a two-phase system one may call the electrode (half-cell) and it can be schematically shown in Fig.2.
The change between the properties of the electronic and ionic conductors must take place over the certain distance (however small). Thus, instead of razor-sharp interface, it is better to think about an interphase region, which is a region of changing properties between

phases. In fact, this region decides about the charge transfer process between the two phases. This charge transfer process is the electrode reaction, and generally it takes place between oxidized and reduced species:

$$Ox + ze = Red \tag{2}$$

which in case of a metal M in contact with the solution containing its ions M^+ can be written as:

$$M^+ + e = M \tag{3}$$

From this electrode reaction results the potential buildup $\Phi_{M+/M}$ at the interface. It is assumed that this reaction is reversible (transfer of charge takes place with the same rate in both directions) and the Laws of thermodynamics can be applied to it. Leaving classification of various electrodes to electrochemists, let's see how this potential can be determined.

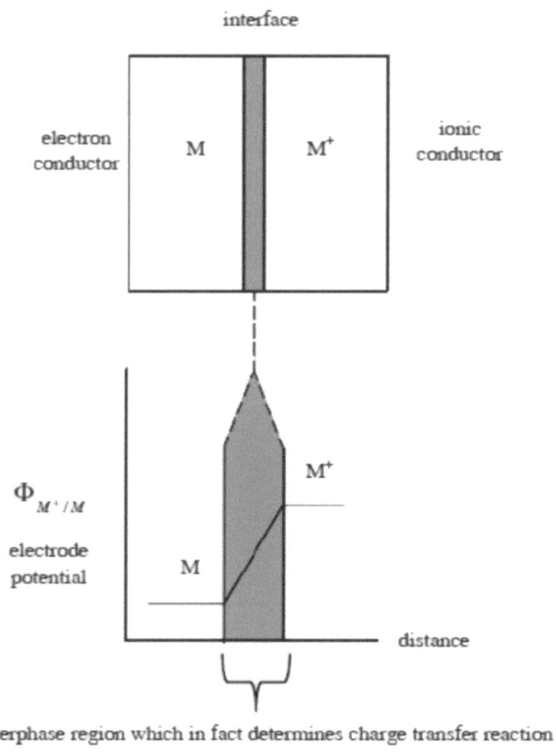

Fig. 2. Scheme of the electrode – electrolyte interface.

The answer to this problem is simple: we need another electrode which must serve as a reference. The single potential cannot be measured, but we can always measure its difference between two half-cells. Thus, the proper construction based on two electrodes yields the source of electric potential difference called electromotive force E (e.m.f.). Such a construction, which is schematically shown in Fig. 3, is called an electrochemical cell.

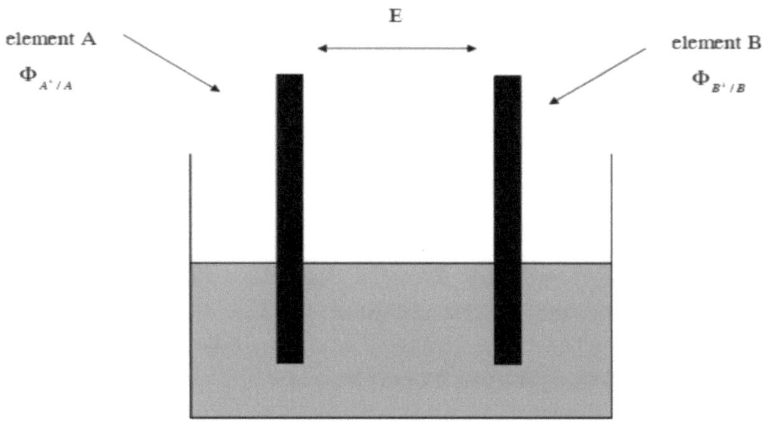

Fig. 3. General scheme of an electrochemical cell.

If this cell is not connected to the external circuit, its internal chemical processes are at equilibrium and do not cause any net flow of charge. Electrochemical energy can be stored. However, if it is connected, then under potential difference the charge must pass from the region of the lower to higher potential. Consequently, to force the charge to flow, work must be done. For any chemical process occurring under constant p and T, the maximum work that can be done by the system is equal to the decrease in its Gibbs free energy:

$$\text{work } W = -\Delta G \tag{4}$$

If this work is electrical one, it equals to the product of charge passed Q = zF and voltage E. For balanced cell reaction, which brought about the transfer of z moles of electrons, this work is given by:

$$W = zFE = -\Delta G \tag{5}$$

where E is cell's electromotive force. From this relationship, the change in Gibbs free energy for the reversible well defined chemical reaction which takes place inside the cell, can be determined as:

$$\Delta G = -zFE \tag{6}$$

where z is number of moles of electrons involved in the process and F is Faraday constant (i.e. the charge of one mole of electrons). Using well-known relations between ΔG, ΔH and ΔS one can express corresponding enthalpy and entropy changes through E vs. T dependence as:

$$\Delta S = zF(\partial E/\partial T)_p \tag{7}$$

and

$$\Delta H = -zFE + zFT(\partial E/\partial T)_p \tag{8}$$

Thus, from measured E vs.T dependence, all thermodynamic functions of the well-defined chemical process taking place inside the cell can be derived.

Under the assumption of chemical equilibrium , eq.6 can be also applied to the electrode reaction. Using the relationship between ΔG and an equilibrium constant K, which is:

$$\Delta G = \Delta G^0 + RT \ln K \quad (9)$$

and combining equations (6) and (9), one can arrive at Nernst's equation:

$$\Phi_{Ox/Red} = \Phi^0_{Ox/Red} - (RT/zF) \ln K \quad (10)$$

in which K is an equilibrium constant written for any electrode reaction in the state of equilibrium (in fact dynamic one). In eq.10, $\Phi_{Ox/Red}$ is an electrode potential, $\Phi^0_{Ox/Red}$ is standard electrode potential (all species taking part in the reaction are at unit activity) and z in a number of moles of electrons taking part in a charge transfer.

Having established all necessary dependencies for electrode potential one can ask how two half-cells can be combined to construct electrochemical cell, and how electromotive force E can be obtained in each case. The general scheme of the cell's classification is shown in Fig.4

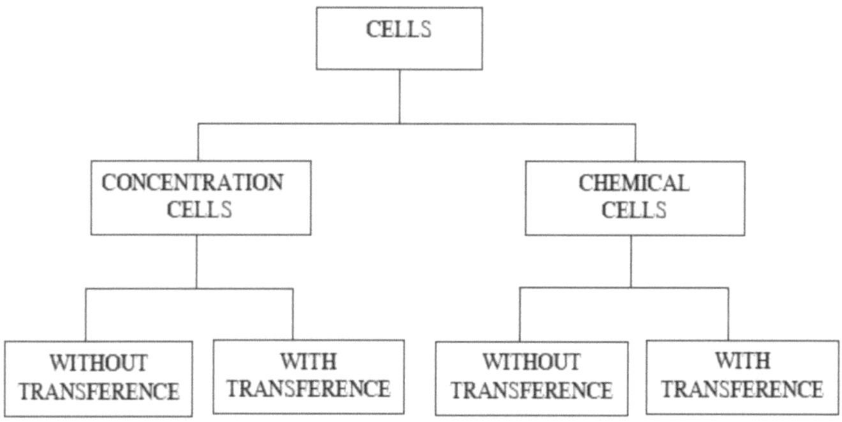

Fig. 4. Classification of the electrochemical cells.

This scheme is based on two characteristic features:

- the nature of the chemical process responsible for the electromotive force production,
- the manner in which the cell is assembled (i.e. where two half-cells are combined into one whole with or without the junction).

Following several simple rules, which say that:

- positive electrode is placed always on the right-hand side of each cell's scheme,
- electrode reactions are always written as reduction reactions,
- electromotive force E for each type of the cell is calculated as the difference of the electrode potentials:

$$E = \Phi_{right} - \Phi_{left} \quad (11)$$

one can analyze each type of the cell construction to see how E is developed, and what kind of thermodynamic information can it deliver.

3. Cells construction

3.1 Chemical cell without transference

We start our considerations from the chemical cells. Schematic representation of this kind of a cell is shown in Fig.5.

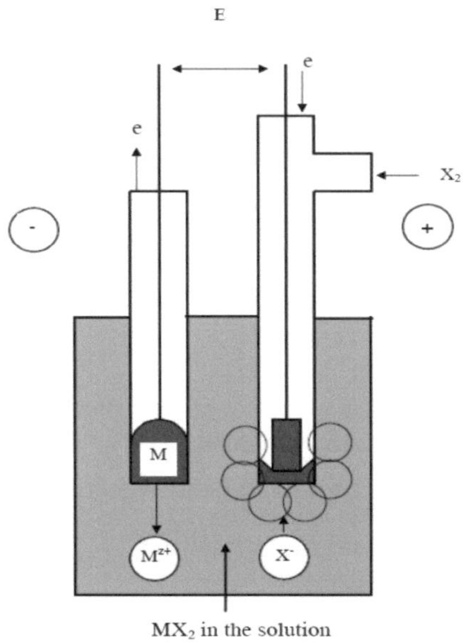

Fig. 5. Scheme of a chemical cell without transference.

Left-hand side electrode consists of the pure metal or alloy, which is immersed into the solution containing its cations (e.g. molten salt). Potential of this electrode written for the reduction reaction :

$$M^{z+} + ze = M \tag{12}$$

is

$$\Phi_{M^{z+}/M} = \Phi^0_{M^{z+}/M} - (RT/zF)\ln(a_M / a_{M^{z+}}) \tag{13}$$

On the right-hand electrode gas X_2 remines in contact with the liquid ionic phase fixing chemical equilibrium of the reaction:

$$(z/2)X_2 + ze = zX^- \tag{14}$$

and establishing the potential :

$$\Phi_{X^-/X_2} = \Phi^0_{X^-/X_2} - (RT/zF)\ln(a^z_{X^-} / p^{z/2}_{X_2}) \tag{15}$$

Consequently, according to the rule mentioned above, the electromotive force of the cell is:

$$E = \Phi_{X^-} - \Phi_{M^{z+}} = E^0 + (RT / zF)\ln(a_M p_{X_2}^{z/2} / a_{M^{z+}} + a_{X^-}^z) \quad (16)$$

Fixing X_2 pressure at the electrode (e.g. p_{X_2} = 1bar), and assuming that $a_{MX_z} = a_{M^{z+}} + a_{X^-}^z$, we have :

$$E = E^0 + (RT/zF)\ln(a_M / a_{MX_z}) \quad (17)$$

It is clear that this type of the cell can be used to measure activities of the components either in metallic or in ionic solution. It is also clear that overall cell reaction is:

$$\underline{M} + (z/2)\, X_2 = \underline{MX_z} \quad (18)$$

and a decrease of Gibbs free energy of this reaction is responsible for the e.m.f. production. The characteristic feature of this cell construction is **the same** liquid electrolyte solution in contact with both electrodes.

3.2 Chemical cell with transference

Another type of chemical cell is so-called Daniell-type cell, in which two dissimilar metals are immersed into two different liquid electrolytes forming two half-cells. To prevent these electrolytes from mixing and consequently, irreversible exchange reaction in the solution, they are separated by the barrier, which however must assure electric contact between both half-cells. The scheme of this cell is shown in Fig.6

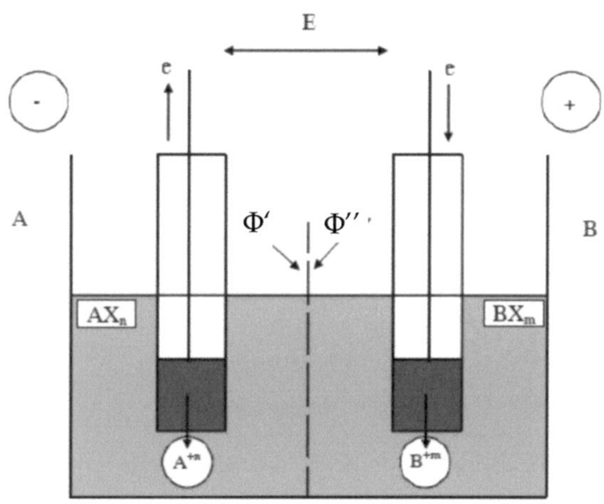

Fig. 6. Scheme of a chemical cell with transference.

The barrier called a junction can be liquid or solid (salt bridge, permeable diaphragm, ion-selective membrane) and can connect the half-cells in a number of different ways . Writing the reduction reaction at the electrodes as:

$$A^{n+} + ne = A \quad (19)$$

$$B^{m+} + me = B \tag{20}$$

and assuming that metals used in electrodes are pure, one can write the expression for electrode potentials:

$$\Phi_{A^{n+}/A} = \Phi^0_{A^{n+}/A} - (RT/nF)\ln(1/a_{A^{n+}}) \tag{21}$$

$$\Phi_{B^{m+}/B} = \Phi^0_{B^{m+}/B} - (RT/mF)\ln(1/a_{B^{m+}}) \tag{22}$$

The e.m.f. of this cell produced by the exchange reaction:

$$mA + nB^{m+} = nB + mA^{n+} \tag{23}$$

is

$$E = E^0 - (RT/zF)\ln(a^m_{A^{n+}}/a^n_{B^{m+}}) \tag{24}$$

where z = nm, activity of metals is equal to one, and expression under logarithm represents equilibrium constant K of the reaction (23). Thus, this cell may provide information about Gibbs free energy change of the exchange reaction at constant temperature, entropy and enthalpy changes can be also obtained if temperature dependence of the e.m.f. is measured. It is not very convenient for high temperature measurements, but can be used successfully while working with aqueous solutions, especially when one half-cell is set as the reference electrode. The characteristic feature of this type of cell is the separation of **two different** electrolytes with the junction assuring electrical contact, but preventing solutions from mixing. Consequently, since two more interfaces in contact with the solution appeared in the cell, there is a hidden potential drop across the junction
$E_{junction} = \Phi'' - \Phi'$ in measured E which not always can be precisely determined. Thus, measurements based on cells with transference may not give as accurate data as chemical cells.

3.3 Concentration cells without transference

If in the same electrolyte solution pure metal and its alloy are submerged, galvanic cell is created. Its scheme is shown in Fig. 7
Two electrode reactions can be written as:

$$M^{z+} + ze = M \tag{25}$$

on the l.h.s , and

$$M^{z+} + ze = \underline{M} \tag{26}$$

on the right, which is more positive.
Corresponding electrode potentials are:

$$\Phi_{M^{z+}/M} = \Phi^0_{M^{z+}/M} - (RT/zF)\ln(1/a_{M^{z+}}) \tag{27}$$

and

$$\Phi_{M^{z+}/\underline{M}} = \Phi^0_{M^{z+}/M} - (RT/zF)\ln(a_{\underline{M}}/a_{M^{z+}}) \tag{28}$$

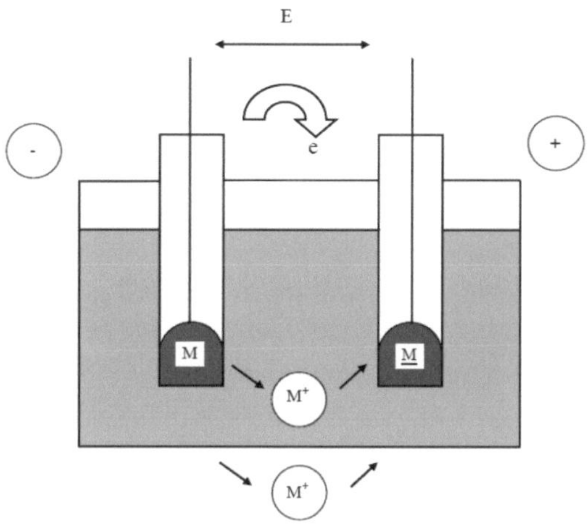

Fig. 7. Scheme of a concentration cell without transference.

The net cell reaction in this case is:

$$M = \underline{M} \tag{29}$$

and E of this cell is generated by the concentration (chemical potential) difference, and has the final form:

$$E = -\ (RT/zF)\ \ln a_M \tag{30}$$

since activity of metal cations is fixed and standard electrode potentials for both electrodes are the same ($E^0 = 0$). From equations (30) and (6) after rearrangement one may obtain:

$$RT \ln a_M = -zFE = \Delta G_m \tag{31}$$

Thus, the free energy change of the transfer process from pure state into the solution can be derived directly from measured E. It is probably the most convenient way to obtain partial function of the alloy component. If the E vs.T dependencies are linear (i.e. E = a+bT), partial entropy and partial enthalpy for the process (29) can be obtained directly for a given composition of the alloy from eqs. (7) and (8):

$$\Delta S_M = zFb \tag{32}$$

$$\Delta H_M = -zFa \tag{33}$$

Again, characteristic feature of this cell is **the same** electrolyte with fixed concentration of M^{z+} ions for both electrodes and the missing junction.

3.4 Concentration cells with transference

The last type of cell is based on the junction which is selectively conducting with one type of an ion. Two versions of this type of cell are schematically shown in Fig. 8. Let's consider the construction given in Fig.8a. Two identical metals (in Fig.8b two identical gaseous species) which are shown are in contact with two electrolyte solutions of two different compositions. These two half-cells are connected through the cation (anion) conducting membrane.

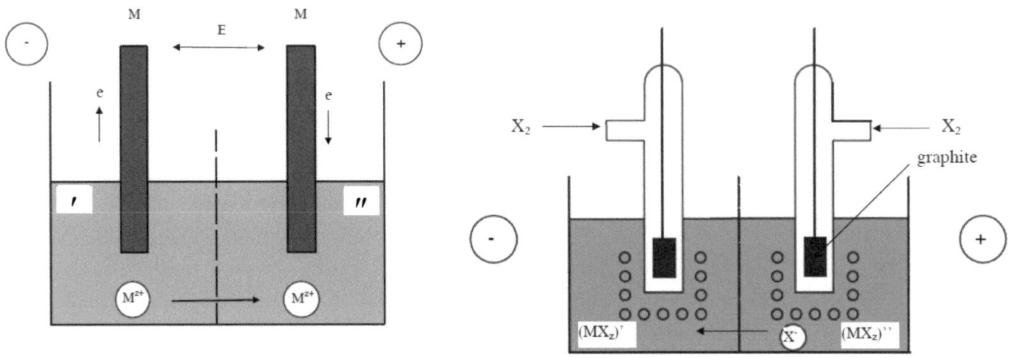

Fig. 8. Scheme of a concentration cell with transference: a) cation conductivity membrane, b) anion conductivity membrane.

If concentration of cations is such that on l.h.s it is higher (a′ > a″) than on the right side, diffusion of cations will proceed and it sets up electric potential difference across the membrane. This potential gradient eventually will stop diffusion and equilibrium is reached. Due to cations migration to the right side, this electrode is more positive. Writing reduction reactions on both electrodes as:

$$(MX)' + e = M + X^- \quad (34)$$

and

$$(MX)'' + e = M + X^- \quad (35)$$

one can express electrode potentials as:

$$\Phi' = \Phi^{0'} - (RT/F)\ln(a_{X^-}/a_{MX'}) \quad (36)$$

$$\Phi'' = \Phi^{0''} - (RT/F)\ln(a_{X^-}/a_{MX''}) \quad (37)$$

Again, since both standard potentials are identical and activity of X^- will not change due to the action of ion- selective membrane which stops their transfer, e.m.f. of this cell (similar reasoning can be given for the cell shown in Fig.8b) is:

$$E = (RT/F)\ln(a_{MX''}/a_{MX'}) \quad (38)$$

If on left side of the cell there is pure MX, then e.m.f. gives directly activity of MX in the solution. However, one must remember that measured E has again an internal contribution from the voltage drop across the junction. Having described principles of cells operation and construction, let's have a brief look at the beginning of the story.

3.4.1 The road to solid electrolytes

Probably, the first-ever type of cells employed to e.m.f. measurements in molten salts was Daniell - type cell employed by Sackur (1918). Relatively easy to construct while working with aqueous solutions, it faced difficulty when temperature of the cell operation was raised. At high temperatures it was used to study molten salts. Concentration cells without transference appeared to be more convenient to study metallic systems. Consequently, this kind of research was initiated by Taylor (1923) and Seltz (1935). Working with cells with transference the main problem connected with this construction was a junction between half-cells and generated junction potential. To avoid interactions between different solutions a number of coupling was tried to assure the electrical contact between the half-cells. While in aqueous solutions the salt bridge was (and still is) the best solution, at high temperature this particular connection brought about same problems. Examples of different half-cell connections tried in the past are shown in Fig.9.

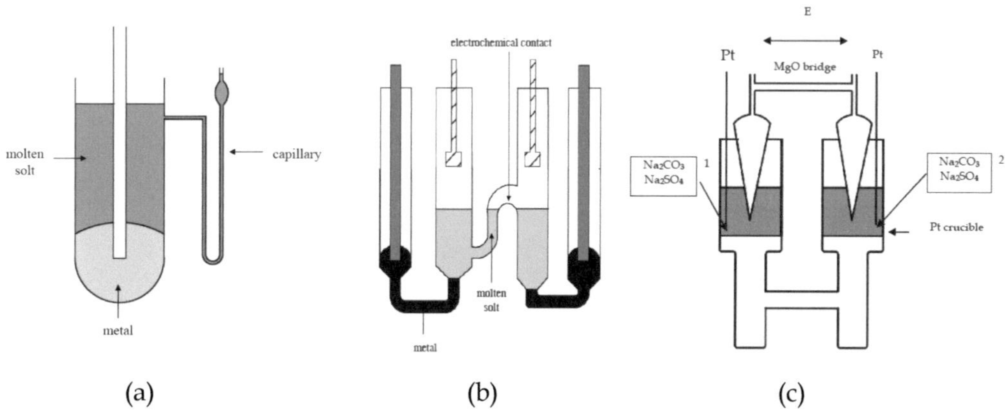

Fig. 9. Different construction of a bridge joining half cells: a) capillary bridge after Lorenz & Michael (1928), b) density difference after Holub et al (1935), c) MgO soaked in molten salt after Flood et al (1952).

In this Figure junctions employed by Lorenz & Michael (1928), Holub et al in 1935 and Flood et al in 1952 are shown. The main problem was to estimate $E_{junction}$ in each case.
In approximately the same time Salstrom (1933) and next Salstrom and Hildebrandt (1930) initiated a series of investigations of molten salts using chemical cells. It was soon realized that proper combination of two chemical cells e.g. $M_1/M_1X/X_2$ and $M_2/M_2X/X_2$ should in principle yield the result equivalent to the result of an exchange reaction completed in the concentration cell $M_1/M_1X//M_2/M_2X$. These findings stimulated both: development of chemical cells which became a source of thermodynamic data for pure substances as well as their solutions, and also employment of concentration cells to provide not only the data for systems for which chemical cells could not be assembled, but also to study the junction potential itself. Its value apparently varied depending on a junction's construction. It appeared to be small for the liquid junction, but as shown by Tamman (1924) who used the cell Ag,AgCl/ glass /$PbCl_2$,Pb, it could be quite significant. Thus, the nature of the junction : liquid or solid, mechanical (frit, gel) or ion-conducting (glass) apparently played significant role in the generated potential drop across the junction. Though experiments with solid substances which played part of electrolytes were already under way, the theory was needed to explain observed discrepancies.

Carl Wagner (1933, 1936) derived the expression for the steady - state ,open – circuit voltage across the scale of solid inorganic compound. Being Walter Schottky's student ,Wagner with his characteristic brightness realized the importance of the defect structure of the solids and its influence on the conductivity. He put forward the theory which gave foundations of the knowledge necessary to design materials, today called solid electrolytes. However, he had to wait almost 25 years to demonstrate applicability of this concept in thermodynamic measurements. In the pioneering work with Kiukkola (1957) they demonstrated that zirconium oxide doped with CaO can be used as the solid electrolyte conducting with oxygen ions. His theory in the simplified version can be explained today in the framework of non-equilibrium thermodynamics.

Let's suppose that through the layer of an inorganic material three species: cations C, anions X and electrons e can migrate. According to Onsager's theory in 1931 their fluxes can be described by the equations:

$$J_C = L_{CC}\,\text{grad}\,\mu_C + L_{CX}\,\text{grad}\,\mu_X + L_{Ce}\,\text{grad}\,\Phi \tag{39}$$

$$J_X = L_{XC}\,\text{grad}\,\mu_C + L_{XX}\,\text{grad}\,\mu_X + L_{Xe}\,\text{grad}\,\Phi \tag{40}$$

$$I_e = L_{eC}\,\text{grad}\,\mu_C + L_{eX}\,\text{grad}\,\mu_X + L_{ee}\,\text{grad}\,\Phi \tag{41}$$

where gradients of chemical potential and electric field play the role of forces responsible for the flow of species and charge through the layer. In the linear regime the matrix of linear coefficients is symmetric i.e. $L_{ij} = L_{ji}$. For the open circuit electric current is zero i.e. $I_e = 0$, and from eq.41 one can obtain:

$$-\,\text{grad}\,\Phi = (L_{eC}/L_{ee})\,\text{grad}\,\mu_C + (L_{eX}/L_{ee})\,\text{grad}\,\mu_X \tag{42}$$

In turn, if there is no gradient of chemical potentials in the system, and the charge flow is caused by the gradient of electric field, taking the ratio of total current and ion fluxes one can arrive at the expressions:

$$J_C / I_e = (L_{Ce} / L_{ee}) \quad \text{and} \quad J_X / I_e = (L_{Xe} / L_{ee}) \tag{43}$$

Introducing transference numbers defined as $I_C/ I_e = t_C = z_C F\, J_C / I_e$ and $I_X / I_e = t_X = z_X F\, J_X/ I_e$, and taking into account reciprocal relations $L_{ij} = L_{ji}$, one can rewrite the equation (42) in the following form:

$$-\,\text{grad}\,\Phi = (t_C / z_C F)\,\text{grad}\,\mu_C + (t_X / z_X F)\,\text{grad}\,\mu_X \tag{44}$$

If inorganic material is conducting only with anions, transference numbers do not depend on chemical potential, and $t_X = 1 - t_e$, then eq.44 can be integrated across the inorganic layer to yield:

$$\Phi'' - \Phi' = E = -\{(1 - t_e)/ z_X F\}\, d\mu_X \tag{45}$$

In the absence of electronic conductivity in the material ($t_e = 0$), and for gaseous species X_2 for which $\mu_X = \mu_X^0 + RT\ln p_{X_2}$, one can arrive at the general formula:

$$E = (RT/z_X F)\ln\left\{p''_{X_2}/p'_{X_2}\right\} \tag{46}$$

Using equation (44) it is also easy to show how liquid junction potential ($E_{junction}$) is generated. Let's assume that instead of the inorganic solid layer, there is the liquid interphase layer between two solutions. In this liquid phase dissociating electrolyte CX yields ions C^{z+} and X^{z-}, which may move independently through the liquid. Then, from (44) one obtains:

$$E_{junction} = - RT/F \ (t^+/z^+) \, d\ln a_+ - RT/F \ (t^-/z^-) \, d\ln a_- \tag{47}$$

Having $z^+ = 1$ and $z^- = -1$, and assuming for the aqueous solution $a_+ = a_- = c_{CX}$, equation (47) can be rearranged and integrated along the layer to yield:

$$E_{junction} = -(RT/F)\,(t^+ - t^-)\ln\{c_2 / c_1\} \tag{48}$$

It is clear that for equal transference numbers (i.e. mobility) the potential drop across the junction is nil.

Since the time of Wagner's paper a number of excellent elaborations was devoted to the solid electrolytes field (Alcock, 1968; Rapp & Shores, 1970; Goto & Pluschkell, 1972; Subbarao, 1980), and we are not going to compete with them. Instead, we'd like to present those areas of research in which, using solid electrolytes, we were able to obtain new data and to small extend we contributed to the extension of the knowledge about thermodynamic properties of high temperature systems.

3.4.2 Oxygen in dilute liquid solutions

To describe the solute element behavior over dilute solution range, free energy interaction coefficients were introduced. However, experimental evidence gathered so far is mainly limited to copper and iron alloys. Working on the review which summarized up to 1988 the data on the solubility of oxygen in liquid metals and alloys (Chang et al, 1988) we found out that there is virtually no information about solute-oxygen interaction in the liquids from which $A_{III}B_V$ semiconducting crystals are grown. The problem is not trival since electrical and optical properties of so-called III-V compounds are influenced by oxygen or water vapor in the growth environment. Oxygen incorporated into crystal brings about a decrease in carrier concentration, photoluminescence efficiency and deterioration of surface morphology. Thus, severe requirements regarding purity of crystals grown from the liquid phase stimulate the need for the data on thermodynamics of solutions containing oxygen dissolved in III-V alloys.

The application of the coulometric titration method to the study of oxygen solubility in liquid metals was first initiated by Alcock and Belford in 1964, and further developed by Ramanarayanan and Rapp in 1972. Our experimental method and the procedure can be described briefly in the following way. Using the electrochemical cell of the type:

$$W, A - B - O \,/\, O^{2-} \,/\, \text{air}, Pt \tag{I}$$

titrations were carried out in the chosen temperature range, which depended on the system studied. In this cell A denotes In and Ga, while B denotes As and P. AB alloys of the chosen composition were prepared in evacuated and sealed silica capsules by melting oxygen free metals with respective MX compounds. Samples were kept at constant temperature for 24 h and then quenched. The tube of the solid electrolyte contained between 2 and 3 g of metallic alloy. A tungsten wire acted as an electric contact with a metal electrode. The outer part of the solid electrolyte tube coated with platinum paste worked as an air reference electrode and was connected to the electric circuit with a platinum wire. The electric circuit contained a potentiostat with a charge meter and digital voltmeter. Purified argon was allowed through the cell just above the surface of the liquid metal. The schematic cell I arrangement is shown in Fig. 10.

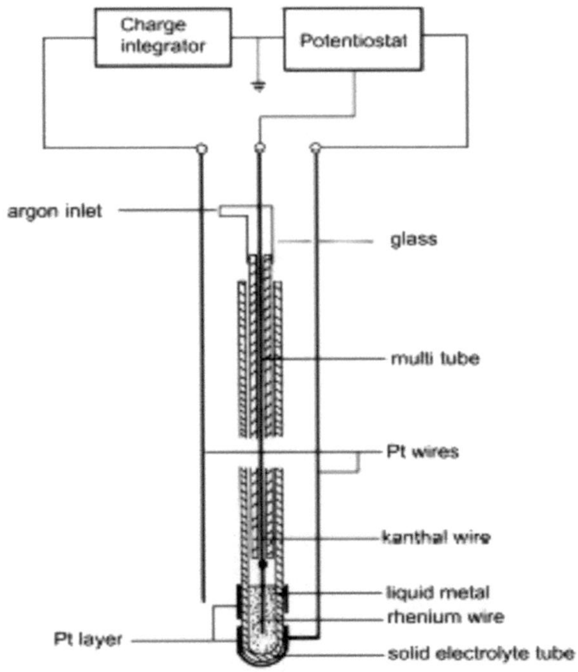

Fig. 10. Scheme of electrochemical cell with air reference electrode by Wypartowicz & Fitzner (1987).

After the equilibrium electromotive force of the cell E_1 had been recorded, the preselected additional potential ΔE was applied by the potentiostat. The resulting current passed through the cell in such a direction that oxygen was pumped out of the alloy. The decrease in oxygen concentration resulted in an increase of the e.m.f. of the cell and a decay of the electric current. The final e.m.f. value E_2 and the electric charge passed Q were recorded. The experimental run was repeated several times at the same temperature, then the temperature was changed.
Activity coefficients of oxygen in liquid alloys were calculated from:

$$f_0 = p_{O_2}^{1/2} / C_{O(1)} \tag{49}$$

where p_{O2} is directly related to the initial e.m.f. through the equation:

$$E_1 = (RT/2F)\ln(0.21/p_{O_2})^{1/2} \tag{50}$$

and the oxygen concentration $C_{O(1)}$ can be obtained from two equations:

$$C_{O(1)} - C_{O(2)} = 100\ (M/W)(Q_{ion}/2F) \tag{51}$$

and

$$E_2 - E_1 = (RT/2F)\ \ln\ (C_{O(1)}/C_{O(2)}) \tag{52}$$

where $E_2 - E_1 = \Delta E$ is an imposed potential difference and the oxygen concentration is expressed in atomic per cent. To obtain eq. 52, Henry's Law was assumed to obey.

In above equations Q_{ion} is the charge corrected by the amount caused by electronic conductivity, M is mass of the sample, and W is atomic weight of the alloy. Experiments were run under following conditions:

- oxygen concentration was kept in the range from 10^{-2} to 10^{-3} atomic percent in order to minimize possible evaporation of oxide species,
- solute concentration varied from 0.01 up to 0.07 X_{As} and 0.05 X_P,
- temperature of experiments (depending on the system) was chosen between 1023 and 1373 K,
- only pump-out experiments were performed.

SYSTEM	ε_0^i	Temperature range	Reference
In-As-O	$31.47 - (39635 / T)$	1023-1123 K	Wypartowicz & Fitzner, 1987
In-P-O	$.589.4 - (803435 / T).$	1100-1200 K	Wypartowicz & Fitzner, 1990
Ga-As-O	$21.10 - (37046 / T)$	1123-1223 K	Wypartowicz & Fitzner, 1988
Ga-P-O	$59.00 - (39635 / T)$	1323-1373 K	Onderka et al, 1991

Table 1. Interaction parameters determined in dilute A-B-O solutions.

Experimental results obtained for four diluted systems, namely In-As-O (Wypartowicz & Fitzner, 1987), Ga-As-O (Wypartowicz & Fitzner, 1988), In-P-O (Wypartowicz & Fitzner, 1990) and Ga-P-O (Onderka et al, 1991) are shown in Table 1. In all cases solute addition decreases activity coefficient of oxygen. This decrease however is the result of a subtle interplay of interactions between A-B atoms on one hand, and A-O and B-O atoms on the other.

3.4.3 Stability of high-temperature ceramic superconductors

The discovery of high temperature ceramic superconductor by Bednorz and Muller (1986) stimulated worldwide research of this type of oxide systems. Research efforts concentrated soon on Y-Ba-Cu-O system, in which $YBa_2Cu_3O_{7-x}$ phase with perovskite structure (designated as Y<123>) showed transition temperature above 90 K (Wu et al, 1987). It was very quickly demonstrated that substitution of yttrium by other rare earth elements is possible. Since the knowledge of phase equilibria in the copper oxide-barium oxide-lanthanide sesquioxide system is necessary for successful synthesis of respective phases, phase stability as a function of temperature and oxygen partial pressure is required to define processing conditions. It appeared however, while working on the Y-Ba-Cu-O system

(Fitzner et al, 1993), that Gibbs free energy of formation of *Ln*<123> phase (*Ln* is a symbol of any lanthanide element) cannot be obtained directly from the e.m.f's produced by the cell with zirconia electrolyte. More sophisticated strategy must be adopted to achieve this aim, and another cell with CaF_2 solid electrolyte must be employed. The measurements require three steps which are schematically shown in Fig. 11.

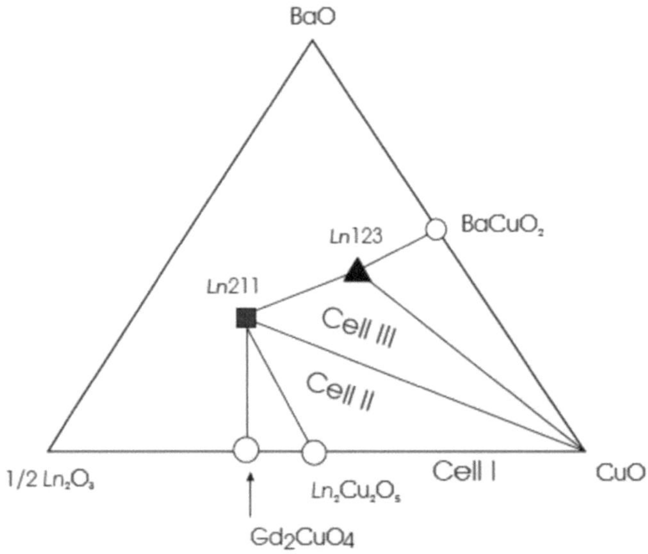

Fig. 11. Scheme of phase equilibria leading of subsequent cells construction.

Cell I is assembled to measure the stability of phases existing in CuO-Ln_2O_3 systems. It is based on zirconia electrolyte and essentially in its construction does not differ much from the cell employed to study liquid dilute solutions (Fig.10). The only substantial difference consists in the working electrode, which in this case is a mixture of oxides. Depending on the size of the lanthanide atom there are two phases which can exist at high oxygen pressure: $CuLn_2O_4$ and $Cu_2Ln_2O_5$. Under decreasing oxygen pressure some of them can be reduced to delafossite - type $CuLnO_2$ .In a series of experiments we determined Gibbs free energy of formation of a number of phases resulting from the reaction between CuO and respective lanthanide oxides. The results of these investigations are summarized in Table 2. Having established ΔG^0 $_{oxides}$ in CuO - Ln_2O_3 systems we could proceed with cell II. It operates with the working electrode CuO-(CuO + Ln_2O_3) phase -Ln_2BaCuO_5 phase (designated as *Ln*<211>). To put this cell into operation, CaF_2 solid electrolyte is required. The application of CaF_2 - type electrolyte in e.m.f. cells employed to the study of oxide systems was described in details by Levitskii in 1978. The cell construction used in our study followed that used by Alcock and Li in 1990. The cell assembly and the whole experimental setup are schematically shown in Fig. 12

Ln	$\Delta G^0_{f,\, oxide} = A + B*T$ [J*mol-1]			Reference
	$Cu_2Ln_2O_5$	$CuLn_2O_4$	$CuLnO_2$	
Lu	39390 – 38.17*T			Przybyło & Fitzner, 1996
Yb	19429 – 22.02*T			Kopyto & Fitzner, 1996
Tm	35275 – 34.09*T			Kopyto & Fitzner, 1996
Er	17427 – 19.61*T			Kopyto & Fitzner, 1996
Ho	18165 – 19.49*T			Kopyto & Fitzner, 1996
Dy	16648 – 17.58*T			Kopyto & Fitzner, 1996
Gd		9562 – 12.29*T		Fitzner, 1990
Eu		885 – 8.67*T	-6640 + 4.02*T	Onderka et al. 1999
Sm		-23500 + 11.06*T	-11250 + 9.64*T	Kopyto et al. 2003
Nd		-28620 + 10.85*T	-18990 + 15.21*T	Kopyto et al. 2003
Y	6670 – 7.05*T		-7265 + 8.31*T	Przybyło & Fitzner, 1995

Table 2. Thermodynamic data obtained for phase existing in Cu-*Ln*-O systems.

The main part of the cell consists of the spring system which was responsible for pressing together working electrode, CaF_2 single crystal and the reference electrode inside the silica tube filled with flowing, synthetic air. The cell was placed in a horizontal resistance furnace which temperature was controlled by an temperature controller. Dry, synthetic air was used to assure that no the traces of moisture can get into the cell compartment. The equilibrium e.m.f. values, which were recorded by an electrometer, were attained in 3 to 10 hours depending on temperature. The platinum lead wires did not show signs of reaction with the electrode pellets after the experiments. Conventional ceramic methods were used to prepare respective phases. Weighed amounts of powders of $BaCO_3$, and CuO were mixed with respective lanthanide oxides, palletized and sintered in the stream of dry oxygen usually at about 1223 K. The e.m.f. measurements were carried out in several cycles of increasing and decreasing temperature. Thermal cycling of the cell produced virtually the same e.m.f. values after reaching constant temperature. The reversibility of the cell performance was also checked by passing the current of 0.1 mA from an external source for about 30 s. The e.m.f. returned to the original values within +/- 1 mV in about 1 – 5 minutes depending on temperature. The whole experimental cycle of the cell operation usually took about one week.
The following electrochemical cells were assembled:

$$CuO + Cu_2Ln_2O_5 + Ln<211> + BaF_2 \,/\, CaF_2 \,/\, CaO + CaF_2 \qquad \text{II}$$

for *Ln* = Yb, Tm, Er, Ho, Dy , and

$$CuLn_2O_4 + Ln<211> + Ln_2O_3 + BaF_2 \,/\, CaF_2 \,/\, CaO + CaF_2 \qquad \text{III}$$

for *Ln* = Gd, Eu, Sm and Nd. Either air or oxygen-argon mixture were flowed continuously through the cell vessel.

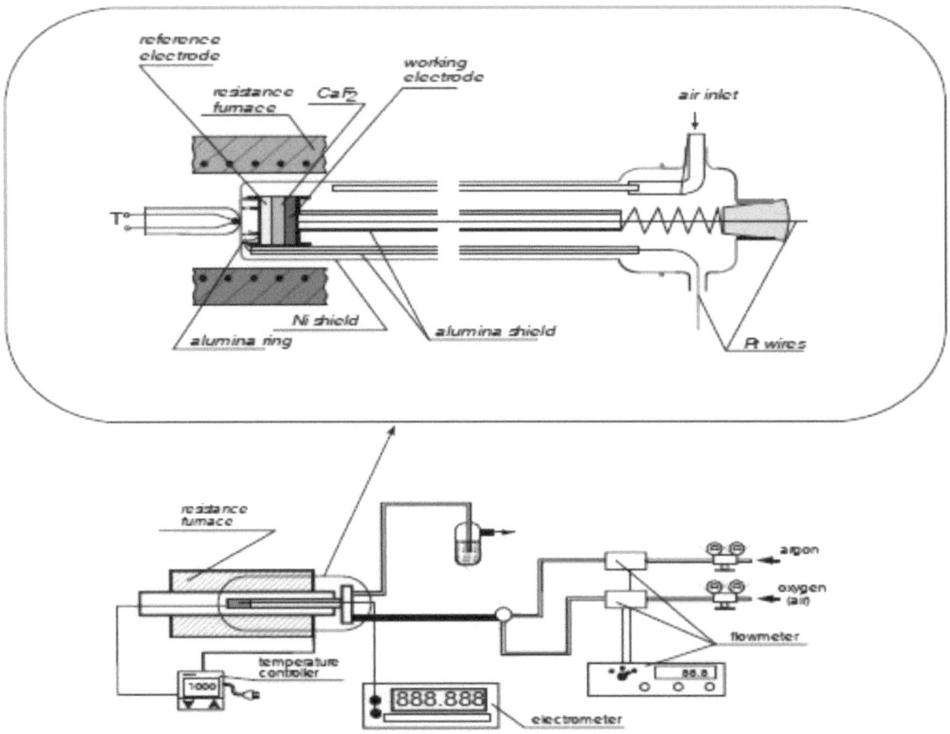

Fig. 12. Scheme diagram of the galvanic cell after Kopyto & Fitzner (1999).

The net reaction for the cell II is:

$$Ln<211> + CuO + CaF_2 = Cu_2Ln_2O_5 + BaF_2 + CaO \tag{53}$$

while for cell III :

$$Ln<211> + CaF_2 = CuLn_2O_4 + BaF_2 + CaO \tag{54}$$

Since mutual solubility between solid phases in the investigated temperature range is small, all components of reactions are essentially in the pure state. Thus, Gibbs free energy change of net cell reactions can be calculated as :

$$\Delta G^0 = -2F\,E \tag{55}$$

If obtained ΔG^0 equations are combined with Gibbs free energy change of exchange reaction between CaO and BaF_2, and also with respective changes of ΔG^0 for reactions of formation of respective double oxides, one can derive Gibbs free energy change for the reaction:

$$CuO + BaO + Ln_2O_3 = Ln<211> \tag{56}$$

Obtained $\Delta G^0_{oxides} = f(T)$ equations are gathered in Table 3
It is the last information needed to set the cell III. It operates with the working electrode: CuO-*Ln*<211> - *Ln*<123>.

Ln	$\Delta G^0_{f, Ln<211>} = A + B * T$ [J*mol-1]	Reference
Yb	-54900 – 4.85*T	Kopyto & Fitzner, 1997
Tm	-51500 – 5.00*T	
Er	-59300 – 1.66*T	
Ho	-52700 – 7.65*T	
Dy	-51200 – 9.06*T	
Gd	-59900 – 1.58*T	
Eu	-82530 + 16.50*T	Przybyło et al., 1996
Sm	-110580 + 41.29*T	Onderka et al, 1999
Nd	-137880 + 65.18*T	Kopyto et al, 2003

Table 3. Thermodynamic data obtained for respective *Ln*<211> phase.

The following electrochemical cells were assembled:

$$CuO + Ln<211> + Ln<123> + BaF_2 \; / \; CaF_2 \; / \; CaO + CaF_2 \qquad (57)$$

for *Ln* = Yb, Tm, Er, Ho, Dy, Gd and Eu.
and their overall cell reaction is:

$$3CaF_2 + 2Ln<123> = 3CaO + 5CuO + 3BaF_2 + Ln<211> + (0.5 - x)\, O_2 \qquad (58)$$

for which the change in Gibbs free energy can be written as:

$$\Delta G^0 = - 6F\, E - (\, 0.5 - x)\, RT \ln p_{O2} \qquad (59)$$

Combining reaction (58) with the exchange reaction between CaO and BaF_2, and with the reaction of formation of *Ln*<211> phases from oxides, one can arrive at the reaction of formation of *Ln*<123> phase from oxides and oxygen:

$$\tfrac{1}{2}\, Ln_2O_3 + 2BaO + 3CuO + (\, 0.25 - x/2)\, O_2 = Ln<123> \qquad (60)$$

Obtained $\Delta G^0 - (\, 0.25 - x/2)\, RT \ln p_{O2} = f(T)$ equations for subsequent *Ln*<123> phases are gathered in Table 4.

Ln	$\Delta G^0 - (0.25 * \frac{1}{2} x) * RT * \ln(p_{O_2}) = A + B * T$ [J*mol-1]	Reference
Yb	-133184 + 16.63*T	Kopyto & Fitzner 1999
Tm	-115143 + 3.96*T	
Er	-122728 + 7.01*T	
Ho	-118424 + 3.23*T	
Dy	-121883 + 7.13*T	
Eu	*-163250 + 44.70*T*	Przybyło et al., 1996

Table 4. Thermodynamic data obtained for respective *Ln*<123> phase.

Dependence of x = f(T) is required under fixed pressure to calculate the correction term. In case of Gd-phase the cell did not work reversibly, probably due to BaF_2*GdF_3 compound formation. In case of Eu, the result should be taken with caution. For elements with the radius smaller than Gd^{3+}, another phase designated as *Ln*<336> appears in the oxide system, and it is a matter of dispute weather this phase is separated from *Ln*<123> by two-phase region or it is simply a solid solution extending from *Ln*<123> phase.

4. Thermodynamic properties of liquid silver alloys

Taking part in two COST projects (531 and MP0602), which were devoted to the search of new lead-free solders, we were involved in the investigations of silver alloys. After all , Poland is a big silver producer and silver alloys seemed to be a reasonable choice. Consequently, using solid oxide galvanic cells with YSZ zirconia electrolyte ($ZrO_2+Y_2O_3$) the thermodynamic properties of a number of the liquid silver alloys were determined. Generally, this kind of galvanic cell can be described by the following schema:

electric contact, working electrode/YSZ electrolyte/reference electrode, electric contact

where the working electrode is metallic alloy of chosen composition mixed with the powder of metal oxide, and the reference electrode is powder mixture of the type Me+MeO or air. First of all in this kind of measurements, the $\Delta G^0_{f,MeO}$ of oxides for each component of the alloy must be known. Next, the activity of this metal which has the most stable oxide can be investigated. Second very important case in these measurements is a lack of an exchange reaction between components of the alloy and metal oxide which is used in this experiment. If Gibbs free energy of formation of two oxides is close to each other the exchange reaction may take place. In this case, the additional experiment must be done to clarify the possibility of an exchange reaction. If these conditions are fulfilled, the chosen silver alloys can be investigated by using e.m.f. method with YSZ zirconia electrolyte. From measured electromotive force E as a function of temperature for alloys of different compositions, the activity of the component of liquid silver alloy can be determined.
The following binary and ternary liquid silver alloys were investigated by using e.m.f. method with YSZ solid electrolyte: Ag-Bi (Gąsior et al., 2003), Ag-Sb (Krzyżak & Fitzner, 2004), Ag-In (Jendrzejczyk & Fitzner, 2005), Ag-Ga (Jendrzejczyk-Handzlik & Fitzner, 2011), Ag-Bi-Pb (Krzyżak et al., 2003), Ag-In- Sb (Jendrzejczyk-Handzlik et al., 2006), and Ag-In-Sn (Jendrzejczyk-Handzlik et al., 2008). In Table 5 galvanic cells which were used during measurements performed of various liquid silver alloys and the range of temperature in which measurements were done are given.

Alloy	Temperature of measurements	Galvanic cell
Ag-Bi	544-1443 K	W, Ag_x-$Bi_{(1-x)}$,$Bi_2O_3/ZrO_2+Y_2O_3$/NiO, Ni, Pt
Ag-Sb	950-1100 K	Re+kanthal, Ag_x-$Sb_{(1-x)}$, $Sb_2O_3/ZrO_2+Y_2O_3$/ air, Pt
Ag-In	950-1273 K	Re+kanthal, Ag_x-$In_{(1-x)}$, $In_2O_3/ZrO_2+Y_2O_3$/ NiO, Ni, Pt
Ag-Ga	1098-1273 K	Re+kanthal, Ag_x-$Ga_{(1-x)}$, $Ga_2O_3/ZrO_2+Y_2O_3$/ FeO, Fe, Pt
Ag-Bi-Pb	848-1123 K	Re+kanthal, Ag_x-Bi_y-$Pb_{(1-x-y)}$, PbO/$ZrO_2+Y_2O_3$/ air, Pt
Ag-In-Sb	973-1200 K	Re+kanthal, Ag_x-In_y-$Sb_{(1-x-y)}$, $In_2O_3/ZrO_2+Y_2O_3$/ NiO, Ni, Pt
Ag-In-Sn	973-1273 K	Re+kanthal, Ag_x-In_y-$Sn_{(1-x-y)}$, $In_2O_3/ZrO_2+Y_2O_3$/ NiO, Ni, Pt

Table 5. Galvanic cells which were used during measurements with silver alloys.

A schematic representation of the cell assembly which was used in measurements of liquid silver alloys is shown in following Fig. 13.

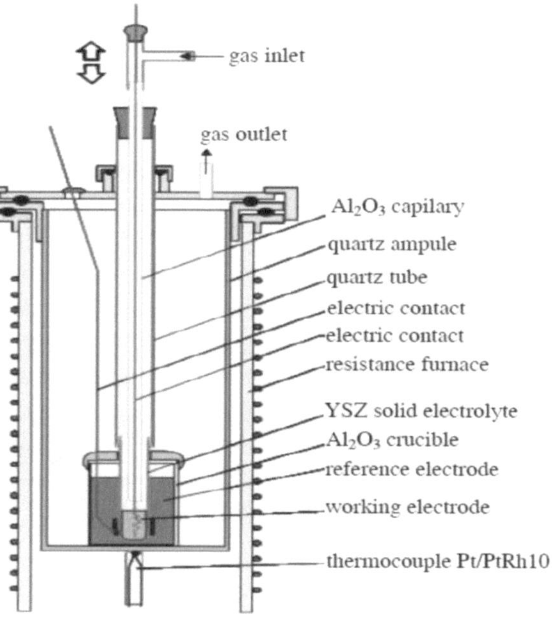

Fig. 13. Galvanic cell assembly after Jendrzejczyk & Fitzner (2005).

The tube of the solid zirconia electrolyte contained a liquid silver alloy of chosen composition. A small amount of pressed metal oxide was placed at the bottom of a YSZ tube. The metallic wires acted as an electric contact with the liquid metal electrode. A solid electrolyte tube was inserted into an alumina crucible filled with reference electrode which was a mixture of metal and oxide powders (in case of measurements with Ag-Sb and Ag-Bi-Pb the reference electrode was air. The construction of the cell was changed and this kind of the cell construction is shown in Fig.10). It was sealed inside the crucible with an alumina cement. Next, the whole cell was placed inside a silica tube, which was suspended on an upper brass head closing the silica tube. The cell was kept in a constant temperature zone of the resistance furnace. The measurement were carried out at increasing and decreasing temperature for several days. To check the reproducibility of the cell performance within the investigated temperature range, measurement were carried out on both heating and cooling cycles. For the reversible overall cell reaction the change Gibbs free energy ΔG was derived from the equation (6). Consequently, by combining equations which describe ΔG and $\Delta G^0_{f,MeO}$ we can obtain equation describing the activity of metallic component in liquid silver alloy. Next all thermodynamic properties of the liquid solution were described in the following manner.

The model for the substitutional solution was used to express the Gibbs free energy of mixing in the following form:

$$G^M = X_1G^0_1 + X_2G^0_2 + \ldots + RT(X_1\ln X_1 + X_2\ln X_2 + \ldots) + G^E \quad (61)$$

where parameters G^0_i denote the free energy of pure component "i" and are taken from the SGTE database (Dinsdale, 1991). The function G^E denotes the excess free energy and for binary system it is described by the Redlich-Kister formula (Redlich & Kister, 1948):

$$G^E_{1,2} = X_1(1-X_1)[L^0_{1,2} + (1-2X_1)^1 L^1_{1,2} + (1-2X_1)^2 L^2_{1,2} + \ldots] \quad (62)$$

For the ternary system, G^E (the excess free energy) is described by the Redlich-Kister-Muggianu polynomial (Muggian et al, 1975):

$$G^E_{1,2,3} = X_1X_2(L^0_{1,2} + L^1_{1,2}(X_1-X_2)) + X_1X_3(L^0_{1,3} + L^1_{1,3}(X_1-X_3)) + X_2X_3(L^0_{2,3} + L^1_{2,3}(X_2-X_3)) + X_1X_2X_3(X_1 L^0_{1,2,3} + X_2 L^1_{1,2,3} + X_3 L^2_{1,2,3}) \quad (63)$$

In above equations the x_i is the mole fraction of the components in an alloy and parameters $L^i_{1,2}$ and $L^i_{1,2,3}$ are linearly dependent on temperature and are given in J/mol*atom.
For subsequent silver alloy shown in Table 5 these parameters were calculated by using experimental data gathered for each system including those which were obtained from our e.m.f. measurements. Parameters $L^i_{1,2}$ and $L^i_{1,2,3}$ for the liquid phase which was obtained during optimization for Ag-Bi, Ag-Sb, Ag-In ,Ag-Ga (Gierlotka & Jendrzejczyk-Handzlik, 2011), Ag-Bi-Pb, Ag-In-Sb and Ag-In-Sn are given in Table 6. Consequently, having these parameters, all thermodynamic properties of the liquid solution were obtained.

Alloy	Parameter [J/mol*atom]
Ag-Bi	$L^0_{Ag,Bi} = 6966.59 - 4.41727 * T$ $L^1_{Ag,Bi} = -5880.40 + 1.54912 * T$ $L^2_{Ag,Bi} = -1942.30$
Ag-Sb	$L^0_{Ag,Sb} = -3619.5 - 8.2962 * T$ $L^1_{Ag,Sb} = -21732.2 + 8.4996 * T$ $L^2_{Ag,Sb} = -6345.2 + 3.2151 * T$
Ag-In	$L^0_{Ag,In} = -13765.1 - 4.803 * T$ $L^1_{Ag,In} = -11431.5 - 0.031 * T$ $L^2_{Ag,In} = -1950.3 + 1.193 * T$
Ag-Ga	$L^0_{Ag,Ga} = -19643.7987 + 63.6589 * T - 8.6202 * T * LN(T)$ $L^1_{Ag,Ga} = -38747.7601 + 140.7162 * T - 16.1115 * T * LN(T)$ $L^2_{Ag,Ga} = -25745.6828 + 140.4553 * T - 17.4166 * T * LN(T)$
Ag-Bi-Pb	$L^0_{Ag,Bi,Pb} = -94764.84 + 125.336 * T$
Ag-In-Sb	$L^0_{Ag,In,Sb} = 86\,430.422 - 42.845 * T$ $L^1_{Ag,In,Sb} = -13\,637.324 - 7.288 * T$ $L^2_{Ag,In,Sb} = -2752.724 + 4.671 * T$
Ag-In-Sn	$L^0_{Ag,In,Sn} = 53\,189.354 - 3.916 * T$ $L^1_{Ag,In,Sn} = 37\,602.644 + 8.536 * T$ $L^2_{Ag,In,Sn} = 1\,044.964 - 39.915 * T$

Table 6. Binary and ternary parameters for the silver alloys.

5. Conclusion

Electrochemical measurements assure versatile and flexible approach to the investigations of high-temperature systems. Not only thermodynamic properties of solid or liquid phases can be determined, but also properly design cell can be used as a sensor to monitor concentration changes in either liquid or gaseous phase. Though at present research efforts made on the development of new sensors shifted towards miniaturization and low temperature working conditions, still high- temperature industrial problems with process and quality control as well as environmental protection require continuous monitoring of the behavior of a number of substances. While the main problem with new sensors development is the transfer of the laboratory cell concept into the device operating under industrial conditions, the basic research is still needed to prepare and understand materials which can be used as potential electrolytes. It is our personal point of view that among variety of choices which can be made future progress will be connected with the investigations of three types of stable solid ionic conductors.

5.1 Ceria – based oxygen conducting solid electrolyte

It is known that substitution of cerium ion with the rare earth trivalent ion on the cerium site in CeO_2 lattice enhances conductivity of ceria, with the highest values obtained for Sm_2O_3 (Inaba & Tagawa, 1996) and Gd_2O_3 additions(Sammes & Du, 2005).

The maximum conductivity of the solid solution is observed between 10 and 20 mol % of Sm_2O_3 and Gd_2O_3 addition, and its value reaches $5*10^{-3}$ S/cm at 500 ^{0}C. Apart from the possible application in fuel cells technology (SOFC), this electrolyte may enable thermodynamic measurements at much lower temperature than zirconia does. Its weakness is the response to relatively high oxygen pressure which causes electronic conductivity in this material at high temperature.

5.2 Lanthanum fluoride solid electrolyte

An incredible fast response of the cell with LaF_3 electrolyte to oxygen pressure changes (about 2 min at room temperature) (Yamazoe et al, 1987) makes it very promising material for various sensor construction. Since the only mobile species in lanthanum fluoride are fluoride ions, the mechanism of this rapid response is not completely known (Fergus, 1997).

It was demonstrated by Alcock and Li in 1990, that measurements of the thermodynamic properties of oxides and sulfides performed with fluoride electrolytes are possible also at high-temperature. Since between SrF_2 and LaF_3 solid solution is formed with the highest melting point about 1835 K at the 70 mole% SrF_2- 30 mole % LaF_3 composition , this solid solution was used as a matrix to form so-called composite electrolyte. In this electrolyte a second phase (in this case either oxide or sulfide) was dispersed and appropriate electrochemical cells were assembled. The results demonstrated that this kind of composite can function satisfactorily and the dispersion of sulfides, carbides or nitrides may lead to the fabrication of electrolytes suitable for sulfur, carbon and nitrogen detection.

5.3 Sulfide ionic conductor

Sulfides are usually less stable and more disordered than oxides. Consequently, it is difficult to find sulfide which exhibit entirely ionic conductivity. The most stable sulfides like MgS, CaS, SrS and BaS exhibit NaCl-type structure with ionic bonds, and have high melting point above 2000 K. Thus, they can be considered as potential solid electrolytes. Nakamura and

Gunji (Nakamura & Gunji, 1980) showed that pure CaS is an intrinsic ionic conductor, however it is not a practical material for the solid electrolyte because of its very small conductivity. In turn, magnesium and strontium sulfides have thermodynamic stability and electrical conductivity comparable to CaS, however at high sulfur pressure (between 10^{-7} and 10^{-4} Pa) p-type conduction appears in these materials.
Recently, Nakamura et al in 1984 obtained highly ionic conductor of sulfides by preparing 95CaS–5Na_2S solid solution. This material exhibits conductivity about two orders in magnitude higher than any other sulfide, which additionally is independent on sulfur pressure. They also demonstrated that the e.m.f. cell Fe, FeS /95CaS-5Na_2S/ Mn, MnS worked reversibly indicating that ionic transference number is equal to one. It is not certain however, which ion is a predominant charge carrier. Though these experiments may be at present only of laboratory interest , still they may lead to the technique enabling to obtain precise thermodynamic data for sulfide systems. Thus, a lot of new problems wait for those who enter this field. We are convinced that this research area is far from being closed. There is still a lot to be discovered and potential applications are enormous and fascinating. We can only hope that at least to certain extent we'll be able to take part in this developing history, and be a witness of new surprises waiting ahead for us.

6. References

Alcock, C. B. (1968). *Electromotive Force Measurements in High-temperature Systems*. Procedings of a symposium held by the Nuffield Research Group, Imperial College, London, England

Alcock, C. B. & Belford, T. N. (1964). Thermodynamics and solubility of oxygen in liquid matals from e.m.f. measurements involving solid electrolytes. *Transactions of the Faraday*, Vol. 60, pp. 822-835

Alcock, C. B. & Li, B. (1990). A Fluoride-based composite electrolyte; *Solid State Ionics*, Vol. 39, pp. *245-249*

Bednorz, J. G. & Muller, K. A. (1986). Possible high T_c superconductivity in the Ba-La-Cu-O system. *Zeischrift für Physic B- Condensed Matter*, Vol. 64, pp. 189-193

Chang, Y. A., Fitzner, K. & Zhang, M. Z. (1988). The solubility of gases in Liquid Metals and Alloys. *Progress in Materials Science,* Vol. 32, No. 2/3, pp. 98-259, ISSN 0079-6425

Dinsdale, A. T. (1991). SGTE Data for Pure Elements. *CALPHAD*, Vol. 15, pp. 317-425

Fergus, J. W. (1997). The application of solid fluoride electrolytes in chemical sensors; *Sensors and ActuatorsB*, Vol. 42, pp. 119-130.

Fitzner, K. (1990). Gibbs free energy of solid phase $CuEu_2O_4$ and $CuEuO_2$. *Thermochimica Acta*, Vol. 171, pp. 123-130

Fitzner, K., Musbah, O., Hsieh, K. C., Zhang, M. X., Chang, A. (1993). Oxygen potentials and phase equilibria of the quatenary Y-Ba-Cu-O system in the region in the involving the $YBa_2Cu_3O_{7-x}$ phase. *Materials Chemistry and Physics*, Vol. 33, pp. 31-37

Flood, H., Forland, T. & Motzfeld, K. (1952). On the oxygen electrode in molten saltz. *Acta Chemica Scandinavica*, Vol. 6, pp. 257-269

Gąsior, W., Pstruś, J., Moser, Z., Krzyżak, A., Fitzner, K. (2003). Surface tension and thermodynamic properties of liquid Ag-Bi solutions. *Journal of Phase Equilibria,* Vol. 24, pp. 40-49

Gierlotka, W. & Jendrzejczyk-Handzlik, D. (2011). Thermodynamic description of the binary Ag–Ga system. *Journal of Alloys and Compounds,* Vol. 509, pp. 38-42

Goto, K. S. & Pluschkell, W. (1972). *Oxygen Concentration cells in Physics of Electrochemistr,* v 2, Ed. J. Hladzik, Academic Press

Holub, L., Neubert, F. & Sauerwald F. (1935). Die prüfung des massenwirkungsgesetzes bei konzentrierten schmelzflüssigen lösungen durch potentialmessungen. *Zeitschrift für Physikalische Chemie,* Vol. 174, pp. 161-198

Hultgren, R., Desai, P.D., Hankin, D. T., Gleiser, M., Kelly, K. K. (1973). Selected Values of the Thermodynamic Properties of Binary Alloys, American Association for Metals

Inaba, H. & Tagawa, H. (1996). Ceria-based solid electrolytes. *Solid State Ionics,* Vol. 83, pp.1

Jendrzejczyk, D. & Fitzner, K. (2005). Thermodynamic properties of liquid silver–indium alloys determined from e.m.f. measurements. *Thermochimica Acta,* Vol. 433, pp. 66-71

Jendrzejczyk-Handzlik, D,; Gierlotka, W & Fitzner, K. (2006). Thermodynamic properties of liquid silver-indium-antimony alloys determined from e.m.f. measurements. *International Journal of Materials Research,* Vol. 97, pp. 1519-1525

Jendrzejczyk-Handzlik, D., Gierlotka, W. & Fitzner, K. (2008). Thermodynamic properties of liquid silver-indium-tin alloys determined from e.m.f. measurements. *International Journal of Materials Research,* Vol. 99, pp. 1213-1221

Jendrzejczyk-Handzlik, D. & Fitzner, K. (2011). Thermodynamic properties of liquid silver–gallium alloys determined from e.m.f. and calorimetric measurements. *The Journal of Chemical Thermodynamics,* Vol. 43, pp. 392-398

Kiukkola, K. & Wagner, C. (1957). Measurements on galvanic cells involving solid electrolytes. *Journal of the Electrochemical Society,* Vol. 104, pp. 379-387

Kopyto, M. & Fitzner, K. (1996)Gibbs energy of formation of $Cu_2Ln_2O_5$ (*Ln* = Yb, Tm, Er, Ho, Dy) and $CuGd_2O_4$ compounds by the e.m.f. method. *Journal of Materials Science,* Vol. 31, pp. 2797-2800

Kopyto, M. & Fitzner, K. (1997). Gibbs free energy of formation of Ln_2CuBaO_5 compounds determined by the EMF method (*Ln* = Yb, Tm, Er, Ho, Dy and Gd). *Journal of Solid State chemistry,* Vol. 134, pp. 85-90

Kopyto, M. & Fitzner, K. (1999). Gibbs free energy of formation of $LnBa_2Cu_3O_{7-x}$ phases determined by the EMF method (*Ln* = Yb, Tm, Er, Ho, Dy). *Journal of Solid State Chemistry,* Vol. 144, pp. 118-124

Kopyto, M., Kowalik, E. & Fitzner, K. (2003). Gibbs free energy of formation of the Nd_2BaCuO_5 phase determined by the e.m.f. method. *The Journal of Chemical Thermodynamics,* Vol. 35, pp. 773-746

Krzyżak, A., Garzeł, G. & Fitzner, K. (2003). Thermodynamic properties of the liquid Ag-Bi-Pb solutions. *Archives of Metallurgy and Materials,* Vol. 48, No. 4, pp. 371-382

Krzyżak, A. & Fitzner, K. (2004). Thermodynamic properties of liquid silver-antimony alloys determined from emf measurements. *Thermochimica Acta*, Vol. 414, pp. 115-120

Levitskii, V. A. (1978). Nekotoryie Pierspiektivy Primyenieniya Metoda E.D.S. so Ftor-ionnym elektrolitom dlya tiermodinamitsheskovo isslyedovaniya tugoplavkich dvoynych okisnych soyediniyenii. *Vestniki Moskovskovo Universiteta, Seria Khimiya*, Vol. 19, pp. 107-126

Lorentz, R. & Michael, F. (1928). Phyrochemische Daniellketten. *Zeitschrift für Physikalische Chemie*, Vol. 137, pp. 1-17

Muggianu, Y. M., Gambino, M. & Bros, L. P. (1975). Enthalpies of formation of liquid alloys bismuth-gallium tin at 723 K. Choice of an analytical representation of integral and partial excess functions of mixing. *The Journal of Chemical Physics*, Vol. 72, pp. 83-88

Nakamura, H. & Gunji, K. (1980). Ionic conductivity of pure solid calcium sulfide. *Transactions of the Japan Institute of Metals*, Vol. 21, pp. 375-382

Nakamura, H., Ogawa, Y., Gunji, K., Kasahara, A. (1984). Ionic and positive hole conductivities of solid magnesium and strontium sulfides. *Transactions of the Japan Institute of Metals*, Vol. 25, pp. 692-697

Nakamura, H., Maiwa, K. & Iwasaki, S. (2006). Ionic conductivity and Transference number of CaS - Na_2S solid solution. *Shigen-to-Sozai*, Vol. 122 pp. 92-97

Onderka, B., Wypartowicz, J. & Fitzner, K. (1991). Interaction between oxygen and phosphorus in liquid gallium. *Archives of Metallurgy and Materials*, Vol. 36, pp. 5-12

Onderka, B., Kopyto, M. & Fitzner, K. (1999). Gibbs free energy of formation of a solid Sm_2CuBaO_5 phase determined by an e.m.f. method. *The Journal of Chemical Thermodynamics*, Vol. 31, pp. 521-536

Onsager, L. (1931). Reciprocal relations in irreversible processes. *Physical Review*, Vol. 37, pp. 405-426

Przybyło, W. & Fitzner, K. (1996). Gibbs free energy of formation of $Cu_2In_2O_5$ by EMF method. *Archives of Metallurgy and Materials*, Vol. 41, pp. 141-147

Przybyło, W. & Fitzner, K. (1996). Gibbs free energy of formation of the solid phases $Cu_2Y_2O_5$ and $CuYO_2$ determined by the EMF method. *Thermochimica Acta*, Vol. 264, pp. 113-123

Przybyło, W., Onderka, B. & Fitzner, K. (1996). Gibas free energy of formation of $Eu_{1+y}Ba_{2-y}Cu_3O_7$ and related phases in the Eu_2O_3-CiO-BaO system. *Journal of Solid State Chemistry*, Vol. 126, pp. 38-43

Ramanarayanan, T. A. & Rapp, R. A. (1972). The diffusivity and solubility of oxygen in liquid tin and solid silver and the diffisivity of oxygen in solid nickel. *Metallurgical and Material Transaction*, Vol. 3, pp. 3239-3246

Rapp, R. A. & Shores, D. A. (1970). *Solid Electrolyte Galvanic Cells in Physicochemical Measurements in metals Research*, v IV. Ed. R. A. Rapp

Redlich, O. & Kister, A. (1948). Algebraic Representation of Thermodynamic Properties and the Classification of Solutions. *Industrial and Engineering Chemistry*, Vol. 40, pp. 345-348

Sackur, O. (1913). Geschmolzene salze als lösungsmittel. III. Mitteilung: Der dissociationsgrad gelöster Salze. *Zeitschrift für Physikalische Chemie,* Vol. 83, pp. 297-314

Salstrom, E. J. & Hildebrand, J. H. (1930). The thermodynamic properties of molten solutions of lead chloride in lead bromide. *Journal of the American Chemical Society,* Vol. 52, pp. 4641-4649

Salstrom, E. J. & Hildebrand, J. H. (1930). The thermodynamic properties of molten solutions of lithium bromide in silver bromide. *Journal of the American Chemical Society,* Vol. 52, pp. 4650-4655

Salstrom, E. J. (1933). The free energy of reactions Involving the fused chlorides and bromides of lead, zinc and silver. *Journal of the American Chemical Society,* Vol. 55, pp. 2426-2428

Seltz, H. (1935). Thermodynamics of solid solutions. II. Deviations from Roult't law. *Journal of the American Chemical Society,* Vol. 57 (March), pp. 391-395

Sammes, N. & Du, Y. (2005). *Intermediate-temperature SOFC Electrolytes, in Fuel Cell Technologies State and Perspectives,* Ed. N. Sammes, Springer pp.19-34

Spencer, P. J. & Kubaschewski, O. (1978). A thermodynamic assessment of the iron-oxygen system. *Calphad,* Vol. 2, pp. 147-167

Subbarao, E. C. (1980). *Electrolytes and Their Applications.* Plenum Press, New York, ISBN 3-306-40389-7

Tammann, Von G. (1924). Die Spannungen der Daniellketten mit flüssigen chloriden und die spannungsreihe der metalle in flüssigen chloriden. *Zeitschrift für Anorganische und Allgemeine Chemie,* Vol. 133, pp. 267-276

Taylor, N. W.(1923). The activities of zinc, cadmium, tin, lead and bismuth in their binary liquid mixtures. *Journal of the American Chemical Society,* Vol. 45, pp. 2865-2890

Turkdogan, E. T. (1980). *Physical Chemistry of High Temperature Technology,* Academic Press New York, ISBN 0-12-704650-X

Wagner, C. (1933). Beitrag zur Theorie des Anlaufvorgangs. *Zeitschrift für Physikalische Chemie-Abteilung B-Chemie der Elementarprozesse Aufbau der Materie,* Vol. 27, pp. 25-41

Wagner, C. (1936). Beitrag zur Theorie des Anlaufvorgangs II. *Zeitschrift für Physikalische Chemie,* Vol. 14, pp. 447-462

Wu, M. K., Ashburn, J. R., Torng, C. J., Hor, P. H., Meng, M. L., Gao, L., Huang, Z. J., Wang, Y. Q., Chu, C. W. (1987). Superconductivity at 93 K in a new mixed-phase Y-Ba-Cu O compound system at ambient pressure. *Physical Review Letters* Vol. 58, pp. 908-910

Wypartowicz, J. & Fitzner, K. (1987). Activity of oxygen in dilute solutions of arsenide in liquid phase with and without gallium additions. *Journal of the Less-Common Metals,* Vol. 128, pp. 91-99

Wypartowicz, J. & Fitzner, K. (1988). Activity of oxygen in dilute solution of arsenic in liquid galium. *Journal of the Less-Common Metals,* Vol. 138, pp. 289-301

Wypartowicz, J & Fitzner, K. (1990). Activity of oxygen in a dilute solution of phosphorus in liquid indium. *Journal of the Less-Common Metals.* Vol. 159, pp. 35-42

Yamazoe, N., Hisamoto, J. & Miura, N. (1987). Potentiometric solid-state oxygen sensor using Lanthanum Fluoride operative at room temperature, *Sensors and Actuators*, Vol. *12*, pp. 415-423.

8

Resonance Analysis of Induced EMF on Coils

Eduard Montgomery Meira Costa
Universidade Federal do Vale do São Francisco
Juazeiro, BA
Brazil

1. Introduction

In all analysis of resonance at RLC circuits, the concept of equality in the inductive reactance and capacitive reactance generates the maximum energy transfer. In the same way, this concept is applied to transformers, due to parasitic capacitances present inter turns (Costa, 2009). Such concept determines explanation of several phenomena of the induced EMF on coils, showing through differential equations and others formalisms of how happens this phenomenon (Costa, 2009a). However, phenomena as secondary energy at Tesla Transformer (that is a pulse transformer (Lord, 1971), where the input energy is a square wave) present some problems that are not fully comprehensible, as the high energy in secondary coil apparently greater then input energy, which was explained in (Costa, 2009b; Costa, 2010) as the sum of responses of the induced EMF at secondary coil. By other side, when analysing these same transformers with input energy as sinusoidal excitation, the induced EMF at resonance appears in the same way, but with smaller gain.
Considering planar coils, several researches are found, applied to microcircuits (Kaware *et al*, 1984; Conway, 2008; Anioin *et al*, 2008; Oshiro, 1987), as well in planar transformers (Oshiro et al, 1989). By other side, several analysis of coils, considering planar coils and ring coils, forming special transformers has been analysed, exciting one (primary) and verifying the other (secondary), in a experimental way, to find new expectations in this area.
When exciting primary coil in these transformers, built with planar coil versus ring coil, called direct system (having the planar coil or the ring coil as primary, and the other as the secondary) or planar coil versus ring coil (called inverted system), the effect of induced EMF at secondary coil presents some common characteristic previously yields studied in circuit theory (as resonance due the RLC characteristic of the transformer). However, with the experimental analysis of the induced EMF in these transformers, when exciting primary coil with square wave and sinusoidal wave, although are found resonance frequencies at frequencies (above 3 MHz) depending of the configuration (number of turns in each coil, and type of coil – planar or ring), gain variations are seen. Also, on resonance frequencies, other phenomena are seen, as high voltage gain in the transformer, even if the primary coil of the transformer presents less turns than the secondary, which contradicts ideal transformer theory.
These cases are analysed here, showing these phenomena, and checking the problem in some mathematical and experimental analysis. The induced EMF from primary coil to secondary coil at transformers in the resonance has shown several properties that generates

new perspectives in electromagnetic theory, as formalisms to develop new type of transformers, others explanations about resonance theory, analysis of energy transfer through resonance on coils, generation of high energy from low power sources, analysis about parasitic capacitances and others characteristics at coils and transformers, and others. This chapter treats of this problem, checking some experimental results, and mathematical formalisms that explain some properties and phenomena that occurs at secondary coil when the resonance frequency is reached in the coupled circuit (transformer), due to induced EMF generated by the excitation of the primary coil with square wave or sinusoidal wave.

2. Experimental methodology and data analysis of Induced EMF at coils

Several experiments were realized in special transformers built in planar coils versus ring coils, to found interesting results about resonance. In all cases, an excitation of square and sinusoidal alternating current were put as entry in primary transformers directly of the wave generator output, to verify the response at secondary coils of the transformers. The analysis of the results shown phenomena at resonance which are analysed here, showing high gains not expected in circuit theory.

2.1 Experimental methodology

In present work were utilized some coils to prepare the transformers where the experiments were realized. These coils are built in copper wire with diameter $d = 2.02 \times 10^{-4}$ m (32 AWG) or $d = 1.80 \times 10^{-4}$ m (36 AWG). Were built several planar coils with diameter of $D = 4.01 \times 10^{-2}$ m, with turn numbers of 20, 50, 200, 500 and 1600, and the ring coils with diameter of $D = 4.65 \times 10^{-2}$ m, with turn numbers of 2, 5, 7, 9, 10, 12, 15, 20, 30 and 50. All coils were built so that their height are $h = 1.8 \times 10^{-4}$ m (case of 20 and 50 turns) and $h = 5 \times 10^{-4}$ m. In this way, the transformers present a planar coil inner ring coil, always based on crossing of the described coils, where this configuration is shown in Fig. 1.

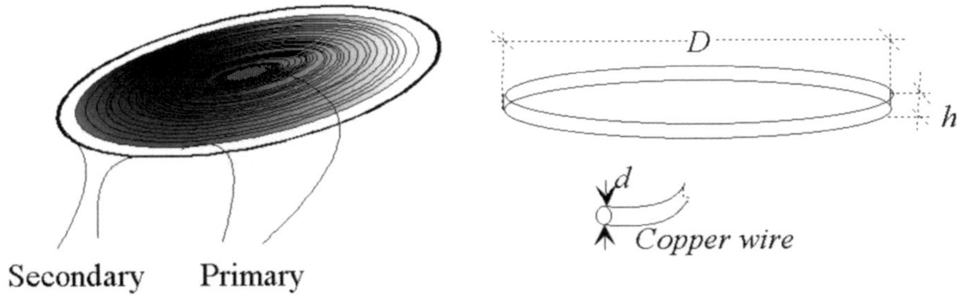

Fig. 1. Basic layer of the studied transformer: Planar coil inner Ring coil.

The measurement of the capacitance between two coils (planar *vs* ring) presented the value of $C_{pr} = 3.1 \times 10^{-11} F$.

The used equipments were: a digital storage oscilloscope Agilent Technologies DSO3202A with passive probe N2862A (input resistance = 10 MΩ and input capacitance ≃12pF), a function generator Rigol DG2021A and a digital multimeter Agilent Technologies U1252A.

Initially, the experiments were realized exciting primary (being the planar coil) with a square wave of 5 V_{pp} (2.5 V_{max}), and frequencies ranging from 1 kHz and 25 MHz, and observing the responses of the secondary open circuit (being the ring coil). Also, with this

same waveform of excitation, the system was inverted, considering ring coil as primary (input of square wave) and planar coil as secondary (output analyzed). *A posteriori*, the excitation was changed by a sinusoidal waveform with the same amplitude, realizing the extraction of the data in direct system and inverted system.
Considering the effects of parasitic capacitances (C_{gi} the parasitic capacitances in relation to a ground, and C_{ci} turn to turn parasitic capacitances, i = 1,2), self (L_i, i = 1,2) and mutual (M) inductances and resistances (r_i, i = 1,2) of coils, the system can be analyzed as equivalent circuit shown in Fig. 2.

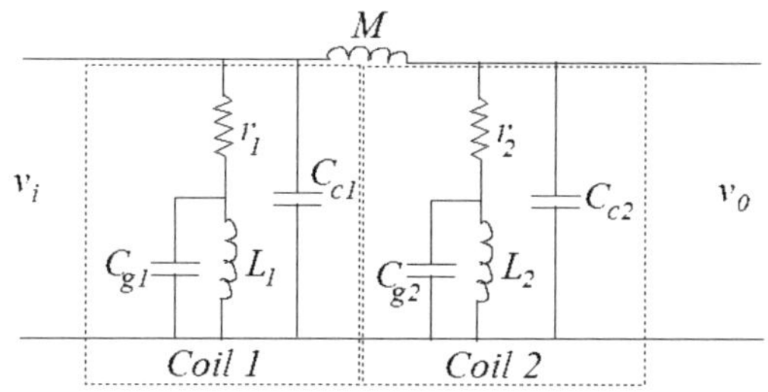

Fig. 2. Equivalent circuit to analysis of the system.

Inductances and mutual inductances of the coils (Hurley and Duffy, 1997; Su, Liu and Hui, 2009) are calculated using procedure presented in (Babic and Akyel, 2000; Babic et al., 2002; Babic and Akyel, 2006; Babic and Akyel, 2008), where some of obtained magnitudes of self inductances are shown in Table 1 as follows:

Turn number	Self inductance (H)
10	4.33×10^{-7}
20	1.85×10^{-6}
30	4.07×10^{-6}
50	1.10×10^{-5}
200	1.67×10^{-4}
500	9.48×10^{-4}
1600	1.07×10^{-2}

Table 1. Computed self inductances for planar coils.

The experimental data obtained with this configuration, *a priori*, were sufficient to determine several effects not common in literature, especially in relation to resonance described as the sum of system responses (when the input is a square waveform), as the high gain that contradicts theory of ideal transformers (in both waveform excitations).

2.1.1 Data analysis for excitation with square wave on direct system

When the primary of the transformer was excited with square wave, the response of the system presented as a second order system, presenting a sinusoidal response with exponential dumping (Costa, 2009d). In this case, considering excitation of the planar coil (as the primary of the transformer), response at secondary is seen in the oscilloscope in time division of 100 μs/div we observe that the system response is verified in accordance with Faraday's law $emf = -d\phi/dt$. This result may be seen in Fig. 3, where the signal of the output is inverted to simplify the observations. In this case, this response is referred to a 200 turns planar coil as primary and a 10 turns ring coil as secondary in input square wave frequency f = 1 kHz. However, when inceasing time division of the oscilloscope for 500 ns/div, in this specific case of the system which generates the response seen in Fig. 3(a), the effects of parasitic capacitances may be observed as attenuated sine wave, as shown in Fig. 3(b). In this case we observe a double sinusoidal (modulated response) with exponential drop. This case is formally observed as effect of the values of the system transfer function, that can be observed only $15 < n_p/n_r < 25$, being n_p is the turn number planar coil and n_r is the turn number ring coil (Costa, 2009; Costa, 2009a).

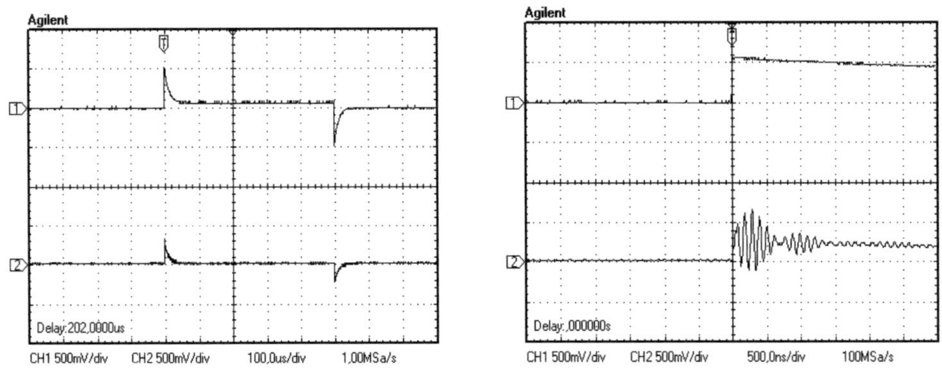

(a) (b)

Fig. 3. (a) Oscilloscope image of input (upper) and output (lower) of the analyzed system with time division 100 μs/div, (b) System response with increasing oscilloscope time division.

When we observe the system responses in other configurations, is observed that the increase of turns in planar coil reduces the lower frequency (that modulates the higher frequency or the main response shown in Fig. 3(b)), as we can see in Fig. 4.

In Fig. 4, we observe that the system response follows equally the rise and the fall of the square wave. Because these effects, when increasing the frequency of square wave applied on primary of the system, we observe that the total system response is presented as the sum of these responses separately. Clearly, the accumulated energy on system (in inductances and parasitic capacitances) is added with the new response when the excitation rises or falls. It is shown in Fig. 5(a), as simulation for 3 attenuated sine wave responses, and at Fig. 5(b) is shown the same effect based on experimental results.

In the case of Fig. 6, we observe that the results have a DC component in response for each rise and fall of the square wave, which is observed in Fig. 7. Consequently this sum of responses is presented as:

$$v_o = \sum_{p=0}^{n} (-1)^p (\alpha \sin(\omega(t-p)) \exp(-b(t-p)) + a) \tag{1}$$

where α is a constant referring to peak response of the sine wave, a is a constant referring to DC level in response, b is a constant referring to exponential attenuation and p is the time when occur each change in the square wave, as we can see in (Costa, 2009c; Costa, 2010a) for some aspects of resonance on coils.

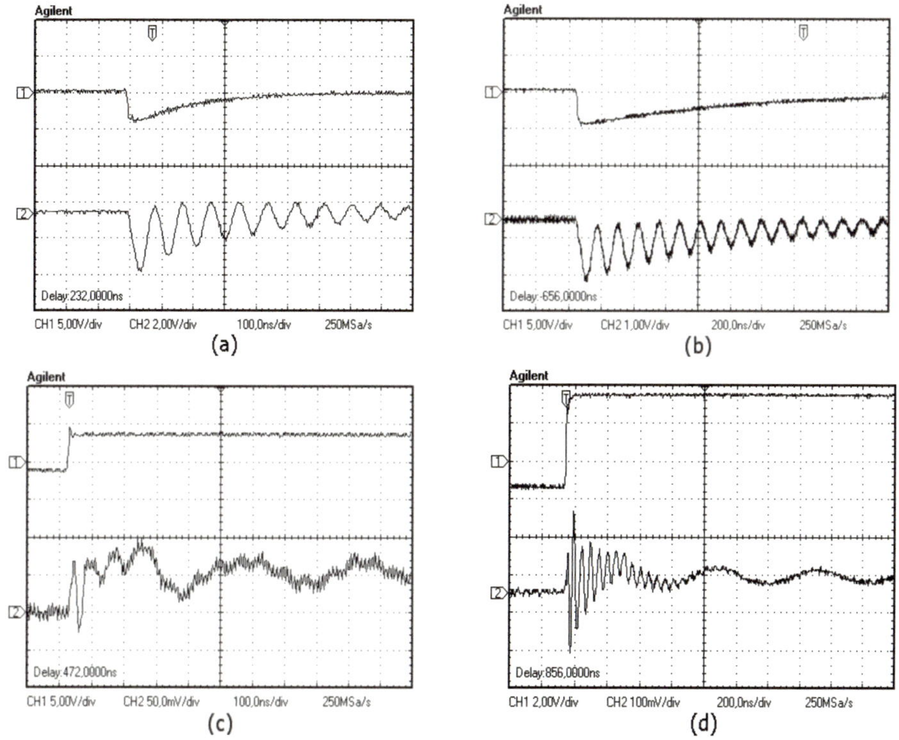

Fig. 4. Responses of the system excited with square wave of f = 1kHz in configurations: (a) 20 turns planar coil *vs* 9 turns ring coil; (b) 50 turns planar coil *vs* 12 turns ring coil; (c) 500 turns planar coil *vs* 7 turns ring coil and (d) 1600 turns planar coil *vs* 5 turns ring coil.

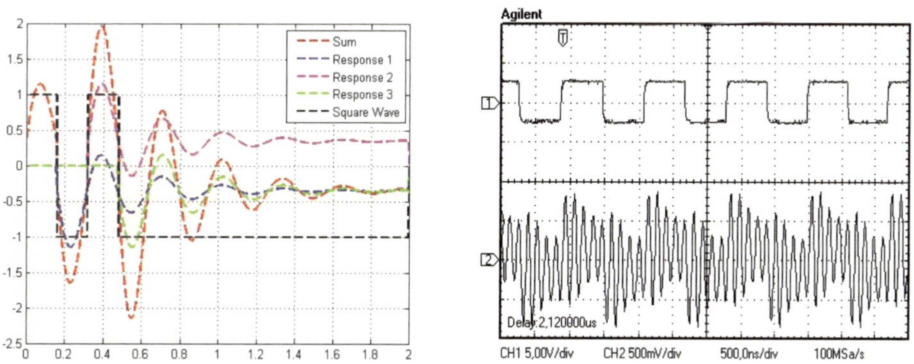

Fig. 5. (a) Simulation showing the sum of attenuated sine wave responses in some rises and falls of the applied square wave on system. (b) Response of the system defined with 200 turns planar coil and 12 turns ring coil, where we observe the sum of individual responses for each rise and fall of the square wave.

Due to the sum of responses, we find that when the responses is in phase with the square wave, i.e., the relation $f_r = f_s/n$, we find a maximum value in output, which refers to the sum of the sine waves in phase and their DC components. In this relation, f_r is the frequency of the main sine wave of the response in each rise and fall of the square wave, f_s the frequency of the square wave and n an integer. Thus, this sum refers to sum of total accumulated energy on coils. However, the resonance only occurs in specific frequencies, when the output is a perfect sine wave. In this case, we can see that effect of resonance is verified when the maximum energy peak is found. This occur in the frequency $f_r = f_s$, which may have values of voltage greater than peak to peak input voltage of the square wave, although the turn ratio of the transformer is lower than 1.

This result may be observed in Fig. 5(a), when the responses are added sequentially, i.e., when $t \rightarrow \infty$, with $f_r = f_s$. In this case, the maximum value of the peak to peak voltage on output is

$$V_{pp_{\max}} = 2\sum_{i=0}^{k}\left(\alpha \exp\left(-b\left(\frac{T(4i+1)}{4}\right)\right) + (-1)^i a\right) \tag{2}$$

where k is the number of cycles of the attenuated sine wave as system response to an input step voltage and T is the period of this sine wave (oscillatory response). Based on this example, we note in all experiments that the found problem due to induced EMF at resonance is that the output is the sum of the responses in each rise and fall of the square wave step voltage. Consequently, it is a result obtained that explains the high voltage of Tesla transformer. These results are shown in Fig. 6, for some configurations of the analyzed system.

In Fig. 6, the obtained data for these configurations are shown in Table 2.

In accordance with these data, we observe that the system response excited with square wave does not follow the common gain of the circuit theory, defined as turn ratio. In other words, in usual circuit theory, the turn ratio determines voltage reduction, but in resonance when is applied square wave as input signal, the response is sinusoidal presenting a visible inversion (high gain defining increased voltage). Thus, for the same data shown in Table 2, data in Table 3 shows the gain of the system in resonance and the expected output value in accordance to circuit theory.

Turn number Planar (n_p)/Ring (n_r)	Turn Ratio n_r/n_p	$v_{pp,max}$ (V)	f (kHz)
20/12	0.6	52.4	8130
50/7	0.14	8.0	13900
200/20	0.1	7.84	4050
500/5	0.01	1.17	18350
1600/30	0.01875	2.02	1570

Table 2. Data of the System Configurations shown in Fig. 8.

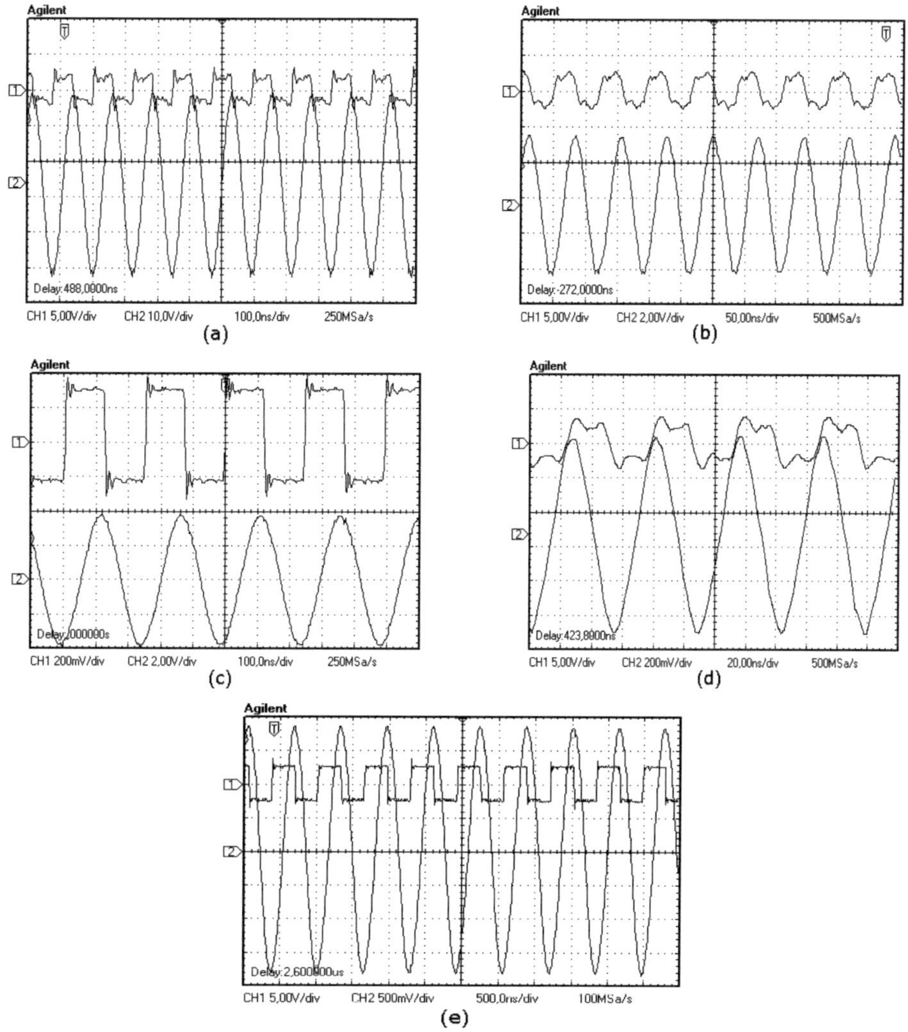

Fig. 6. Resonance of the systems in configurations: (a) 20 turns planar coil *vs* 12 turns ring coil on f = 8130 kHz; (b) 50 turns planar coil *vs* 7 turns ring coil on f = 13900 kHz; (c) 200 turns planar coil *vs* 20 turns ring coil on f = 4050 kHz; (d) 500 turns planar coil *vs* 5 turns ring coil on f = 18350 kHz and (e) 1600 turns planar coil *vs* 30 turns ring coil on f = 1570 kHz.

Turn number Planar/Ring	$v_{pp,max}$ (V)	v_0/v_i	v_{pp} - expected value of circuit theory (V)
20/12	52.4	10.48	3.0
50/7	8.0	1.6	0.7
200/20	7.84	1.568	0.5
500/5	1.17	0.234	0.05
1600/30	2.02	0.404	0.09375

Table 3. Ratio Output/Input and Expected output for Data at Table 2.

These effects are clearly visible when using Equations (1) and (2), which show why the system at resonance can get high energy.
The same effect is observed for the inverted system, i.e., when ring coil is the primary of the transformer and the planar coil is the secondary. This case is presented in the next section.

2.1.2 Data analysis for excitation with square wave with inverted system

Considering the inversion of the system, i.e., ring coil as primary and planar coil as secondary, the response appears similarly to initial configuration. But due to the inversion of the values (parasitic capacitances, self inductances and resistances in the equivalent system shown in Fig. 2, and consequently changes in value of mutual inductance (Babic and Akyel, 2000; Babic et al., 2002; Babic and Akyel, 2006; Babic and Akyel, 2008) changes in transfer function are made, such that the output presents features similar to the cases where n_p/n_r >25. In these cases, the inversion of the values in transfer function also generates a lower frequency on oscillatory response. Consequently, the system response presents resonance in lower frequencies than the initial configuration, as we see in (Costa, 2009d).
Observing Fig. 7, we see the system responses for some configurations when the input signal is a low frequency square wave (similarly to input step voltage). In this figure, we clearly observe that the frequencies are lower than frequencies of system response in initial configuration. Also, we observe that the input square wave is presented with effects of *RL* circuit, due to passive probe of oscilloscope be in parallel to primary coil (Babic and Akyel, 2000; Babic et al., 2002; Babic and Akyel, 2006; Babic and Akyel, 2008).
When the frequency is increased, the same effect of sum of responses to each rise and fall of the square wave defined in Equation (1) is observed, as shown in Fig. 8. In the same way, when the relation $f_r = f_s/n$ is verified, the output voltage reaches the maximum value, although this response is not a perfect sine wave.
However, in accordance to Equations (1) and (2), when the relation $f_r = f_s$ is verified, the resonance occurs, and the output reaches the maximum value with a perfect sine wave. Since that the frequencies of the responses are lower, the resonance occurs in low frequencies of the square wave, in comparison with the initial configuration. Some results of this case are shown in Fig. 9.
We observe in this case, that the output voltage (v_{pp}) is greater than the initial configuration. Clearly, this effect is observed because two components are considered: the turn ratio (effect of the transformer, as circuit theory) and the sum of the sinusoidal responses as (1). Consequently, the resonance output voltage is greater than the effect of the transformer alone.
For configurations shown in Fig. 7, 8 and 9, the maximum values of the output voltage are shown in Table 4, with their respective turn ratio and resonance frequency.
However, although in this case occurs an effect of the turn ratio (transformer), in accordance to results shown in Table 4, this effect defines that this is not always right, as in the case of the configurations of 2 turns ring coil *vs* 1600 turns planar coil, 2 turns ring coil *vs* 50 turns planar coil, 5 turns ring coil *vs* 500 turns planar coil and others with turn ratio greater than 100. It is due to impedance of the circuit, which eliminates various sinusoidal components of the input square wave, reducing total value on output.
In the realized measurements with all coils in the initial configuration and inverted system, we can see the behavior of the output voltage when varying turn number of the coils (ring and planar) in Fig. 10, for input square wave of 5 V peak to peak. With this Fig. 10 we can generate a direct comparison for both cases worked, verifying the gain.

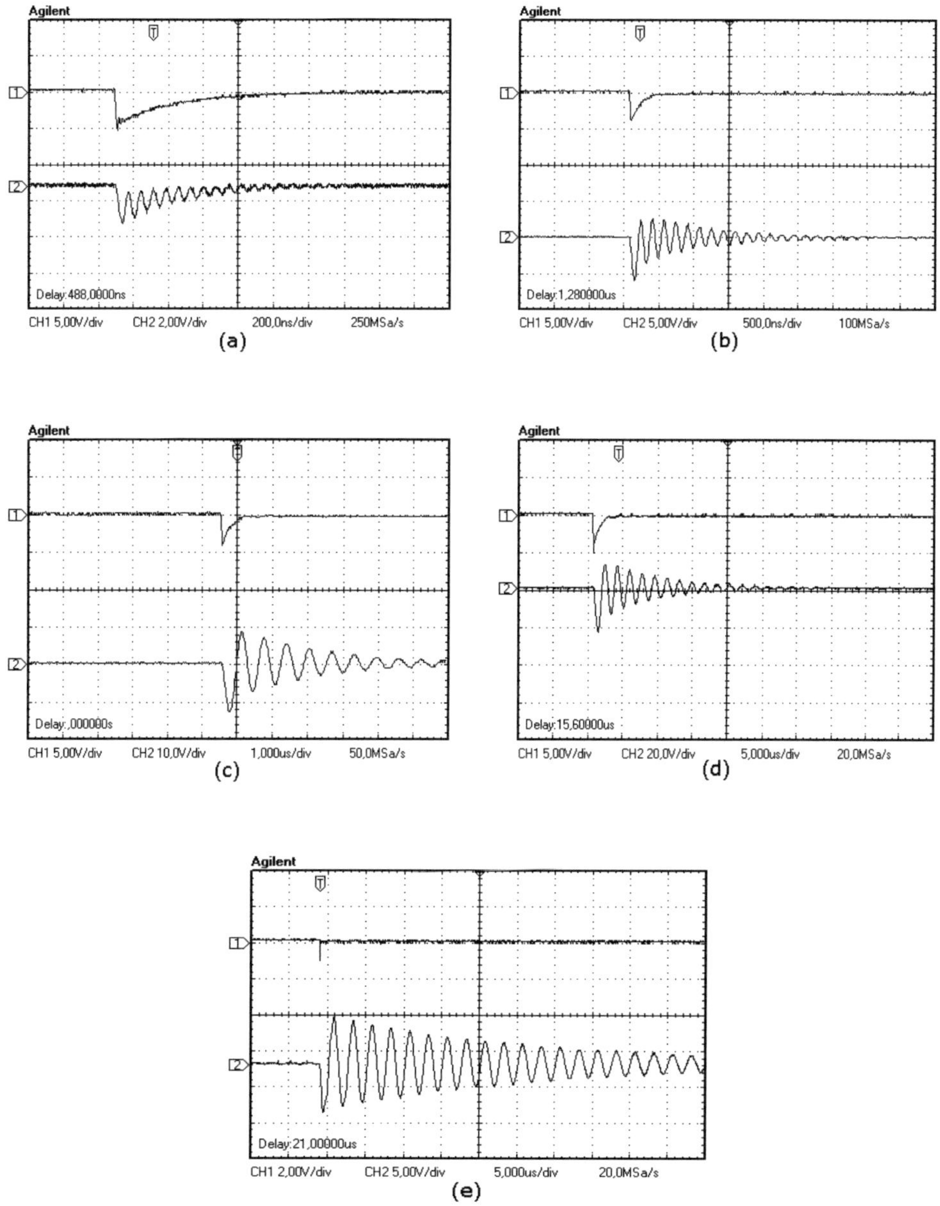

Fig. 7. Some responses of the inverted system: (a) 12 turns ring coil *vs* 20 turns planar coil; (b) 7 turns ring coil *vs* 50 turns planar coil; (c) 9 turns ring coil *vs* 200 turns planar coil; (d) 20 turns ring coil *vs* 500 turns planar coil and (e) 2 turns ring coil *vs* 1600 turns planar coil.

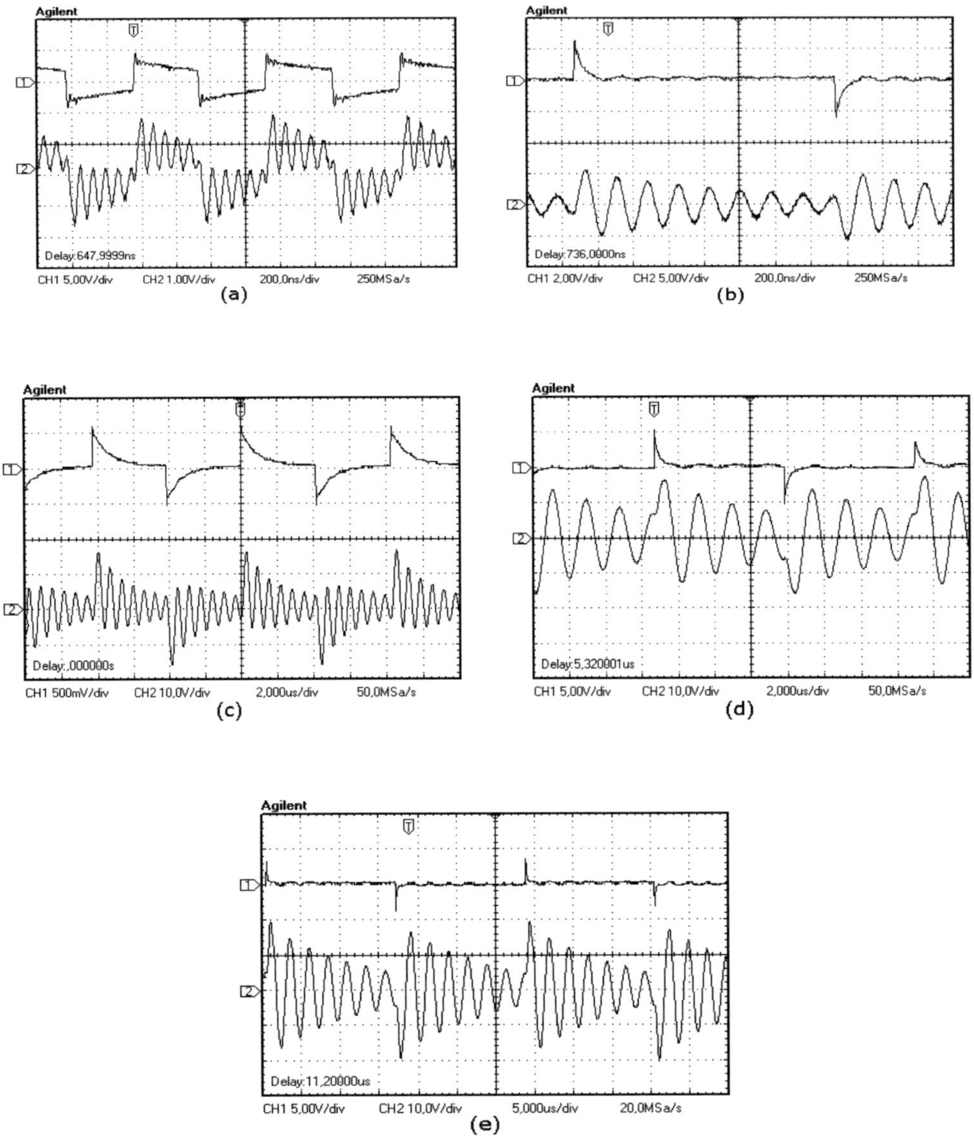

Fig. 8. Sum of responses on inverted system in configurations: (a) 15 turns ring coil *vs* 20 turns planar coil; (b) 2 turns ring coil *vs* 50 turns planar coil; (c) 20 turns ring coil *vs* 200 turns planar coil; (d) 10 turns ring coil *vs* 500 turns planar coil and (e) 7 turns ring coil *vs* 1600 turns planar coil.

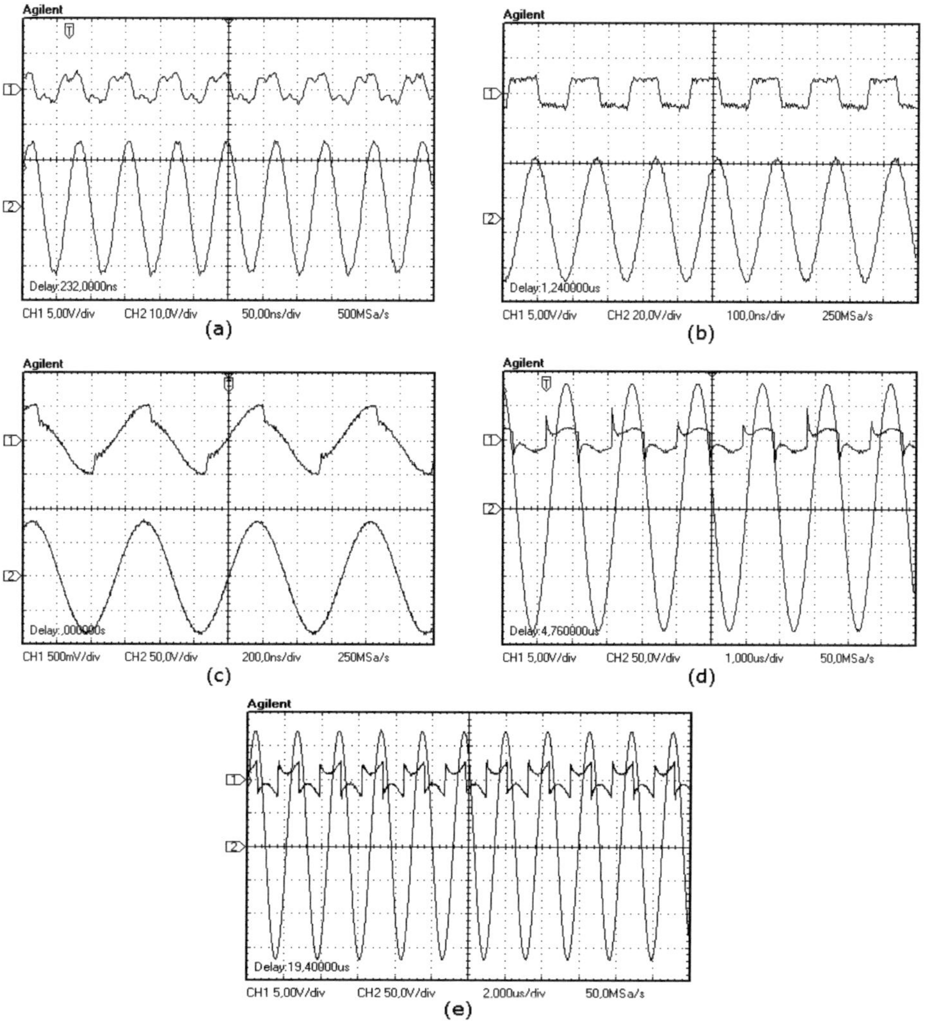

Fig. 9. Resonance in some inverted systems: (a) 7 turns ring coil *vs* 20 turns planar coil; (b) 15 turns ring coil *vs* 50 turns planar coil; (c) 12 turns ring coil *vs* 200 turns planar coil; (d) 5 turns ring coil *vs* 500 turns planar coil and (e) 30 turns ring coil *vs* 1600 turns planar coil.

In case of the initial configuration (direct system), the gain obtained is increased follows in accordance to turn numbers of the both coils. In the case of the inverted system, the gain is decreasing to turn number ring coil, and reaches the maximum peak voltage in configurations defined as low turn number in ring coil and high turn number in planar coil (in the obtained experimental results this value is 5 turn number ring coil). The obtained values for turn number in ring coil lower than 5 is decreasing, when it is crossed with turn number higher than 200 in planar coils. Naturally, this effect is verified as being the variation of the values of parasitic capacitances, self-inductances and mutual inductance (Babic and Akyel, 2000; Babic et al., 2002; Babic and Akyel, 2006; Babic and Akyel, 2008), since that these planar coils are built in more than one layer in the same disk diameter.

Turn number Ring (n_r)/ Planar (n_p)	Turn Ratio n_p/n_r	$v_{pp,max}$ (V)	f (kHz)
12/20	1.667	21.4	13900
7/50	7.143	96	5920
9/200	22.222	170	1573
20/500	25.0	328	560
2/1600	800.0	322	402
15/20	1.333	17.6	13900
2/50	25.0	99.2	5700
20/200	10.0	174	1580
10/500	50.0	336	540
7/1600	228.571	400	407
7/20	2.857	38.8	14000
15/50	3.333	72.8	5770
12/200	16.667	166	1520
5/500	100.0	362	530
30/1600	53.333	342	442

Table 4. Results to Inverted System in Configurations of Fig. 7, 8 and 9.

Also, other effect observed in inverted system is the output peak voltage for low turn number in ring coil. When the turn number in planar coil is increased, considering 5 turn number in ring coil, the graph seen in Fig. 10(b) increases quickly, showing a better relationship to maximum response in resonance.

Because this relationship between coils, higher sinusoidal voltage is obtained, showing important results between self inductances and others parameters evaluated on system to generate high voltages on air-cored transformers, as well how we can built small pulse transformers [28] with high voltage output based on planar coil inner ring coil and others applications for energy transfer.

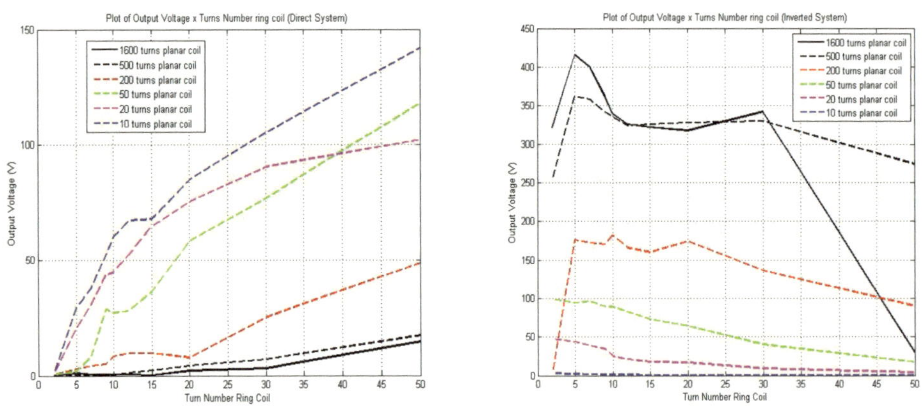

(a) (b)

Fig. 10. Output variations according to turn number coils: (a) Initial configuration; (b) Inverted System.

Finally, we observe that these results, in both cases (initial configuration or direct system, and inverted system) show important effects on resonance in pulsed systems, when they involve coils, which may be used to analysis on electromagnetic interference and other problems of power electronics, pulse transformer and computational systems.

2.1.3 Data analysis for excitation with sine wave in direct system and comparison with square wave excitation

When analysing the response of the system when excitation is a sinusoidal waveform, the result presented is a sinusoidal wave with phase variation. But, when the resonance is reached, high gain is noted, however slightly larger than the response of the excitation with square wave. In both cases, the induced EMF at resonance generates the phenomenon of that the coupled circuit theory (transformer) and circuit theory is not satisfied, since that the high gain do not satisfies the ideal gain of the transformers.

Considering direct system, the response at low frequencies (from 1 kHz to 20 ~ 50 kHz) are generally noise, as shown in Fig. 11. It is due to effect of inductances, mutual inductances, and parasitic capacitances, which determines a filter which cut these frequencies. For frequencies above 50 kHz, response of the system appears as lagged sine wave, as we can see in Fig. 12.

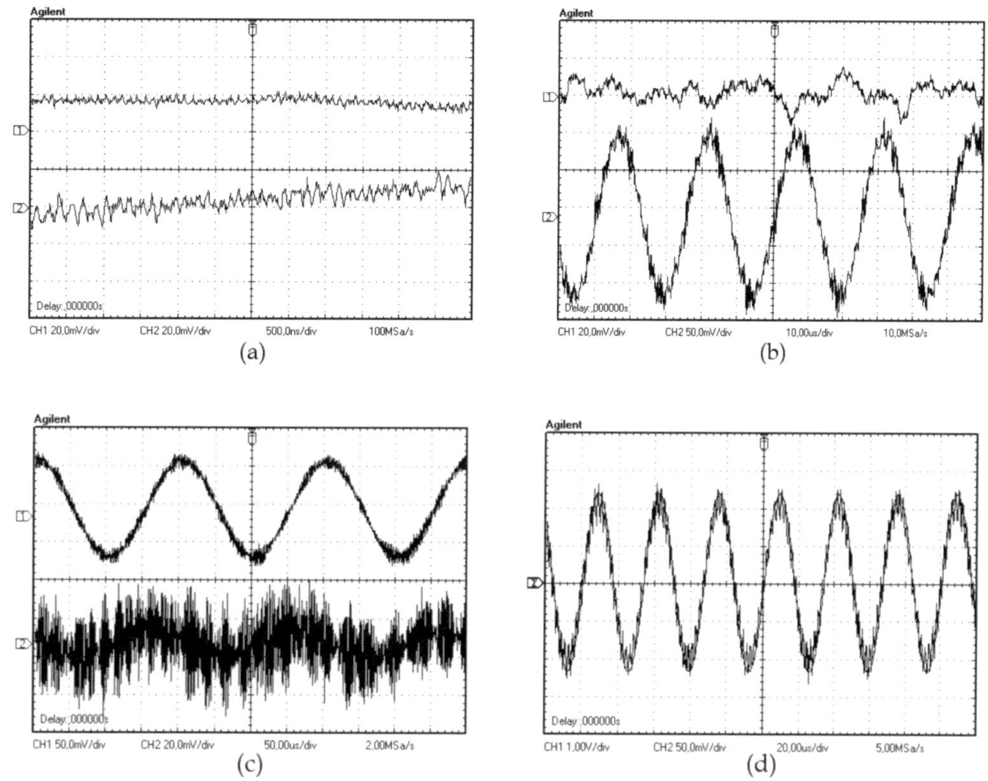

Fig. 11. (a) 10 turns planar coils versus 15 turns ring coil at 30 kHz; (b) 20 turns planar coils versus 30 turns ring coil at 40 kHz; (c) 30 turns planar coils versus 10 turns ring coil at 5 kHz; (d) 200 turns planar coils versus 10 turns ring coil at 50 kHz;

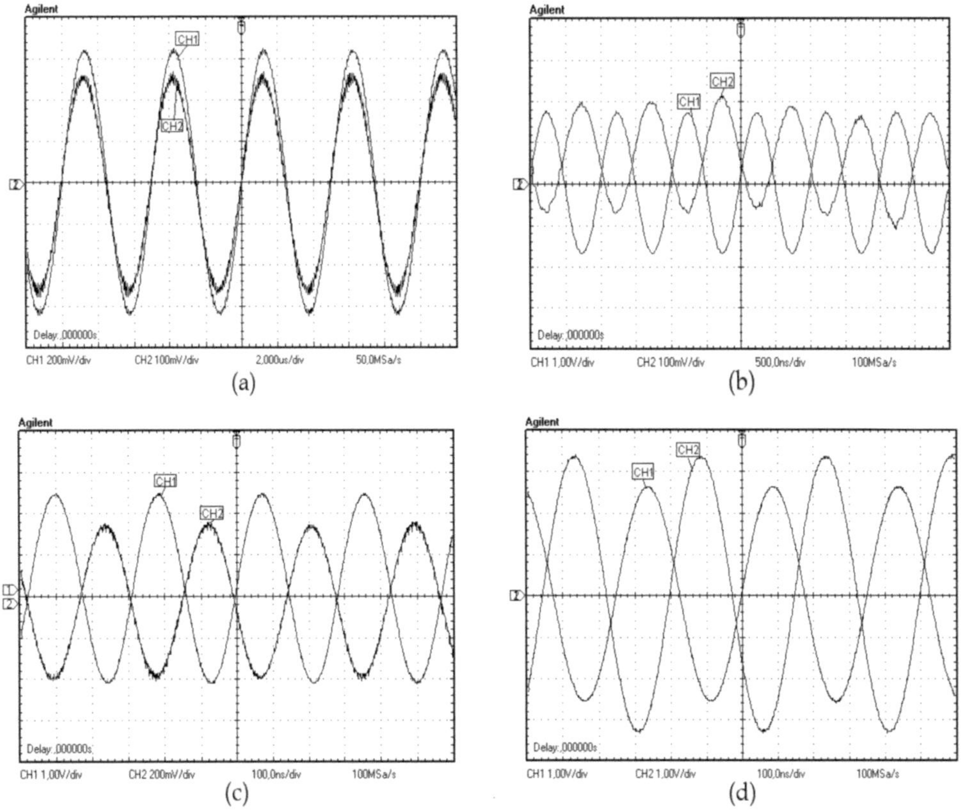

Fig. 12. (a) 20 turns planar coils versus 10 turns ring coil at 200 kHz; (b) 30 turns planar coils versus 15 turns ring coil at 1000 kHz; (c) 50 turns planar coils versus 50 turns ring coil at 3500 kHz; (d) 200 turns planar coils versus 30 turns ring coil at 2850 kHz.

Analysing the problem of these frequencies, we can see that the response of the system is:

$$v_o = A(\omega)\sin(\omega t + \phi(\omega)) \tag{3}$$

where $A(\omega)$ is the amplitude and $\phi(\omega)$ is the phase, which both depends upon frequency. Amplitude varies with inductances, mutual inductances, resistances and parasitic capacitances of the system, presenting similar graph as system excited with square wave, but with fewer resonance peaks, as view in Fig. 13, which compares output of the system in some cases excited by sine waves and square waves. The number of resonance peaks in output of the system when escited by square waves is due to components of the Fourier series that passes at filter, generating several resonance peaks with increased amplitudes, as frequency increases,which similarly we find in (Cheng, 2006, Huang et al, 2007) analysis of problems involving harmonic analysis.

In the case of sine wave excitation, resonance responses are found as higher output as excitation with square waves. When considering the sinusoidal excitation, the maximum gain are presented at Table 5, where are seen the gains of some experimented systems, considering ring coil as primary, and in Table 6 are seen obtained ratio of these two gains for direct system (ring coil as primary), where are seen that in resonance with sine wave excitation, the gain is higher than square wave excitation.

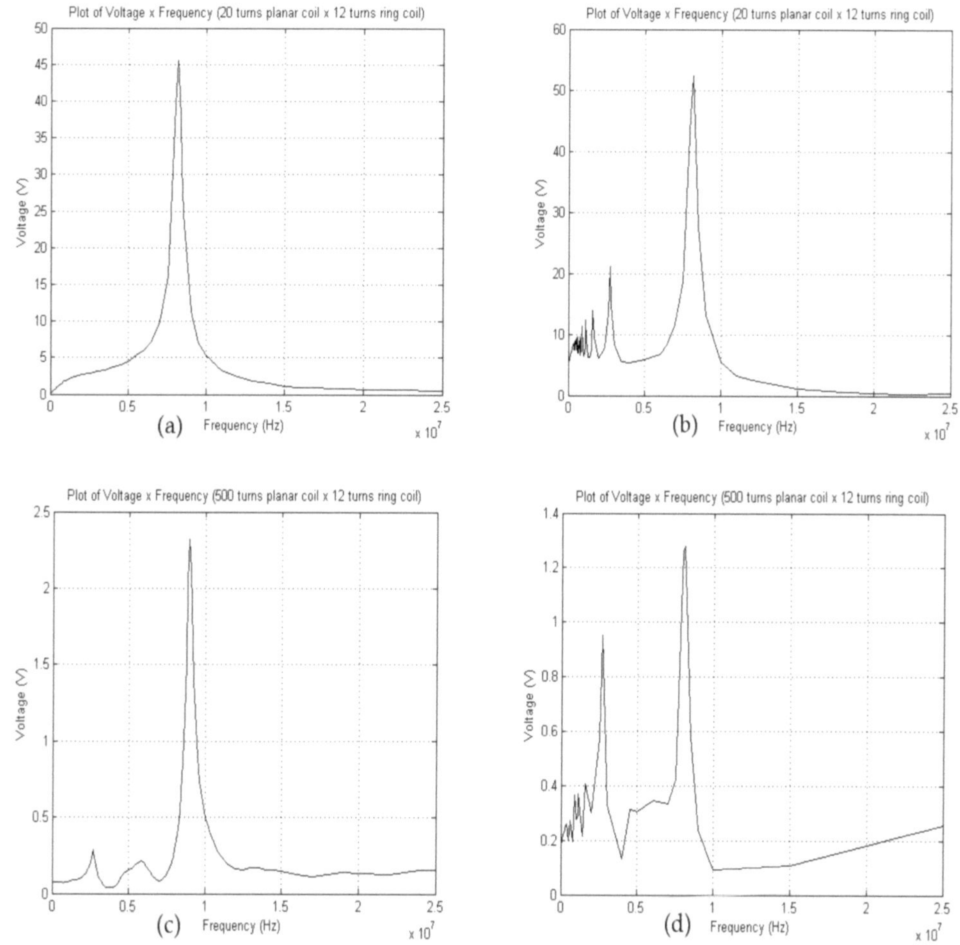

Fig. 13. Graphs of voltage versus frequency of the system built in 20 turns planar coil versus ring coil excited by: (a) sine wave; (b) square wave; and system built in 500 turns planar coil versus 12 turns ring coil excited by (c) sine wave; (d) square wave.

Coil	10	20	50	200	500
10	0.14	3.80	12.40	28.40	50.00
12	0. 46	1.84	13.36	13.04	50.80
20	0.22	1.26	11.88	22.80	51.60
30	0.10	1.47	6.12	19.20	5.04
50	1.98	2.20	6.04	5.04	0.16

Table 5. Gains of transformers with sinusoidal excitation (direct system): columns with planar coil; rows with ring coil.

Coil	10	20	50	200	500
10	0.39	1.58	1.46	1.61	1.49
12	1.68	0.77	1.67	0.82	1.57
20	0.93	0.83	1.95	1.36	1.57
30	1.74	1.84	1.59	1.50	0.15
50	3.17	5.29	3.68	0.60	0.01

Table 6. Ratio of the gain transformers with sinusoidal (sin) and square wave (sw) excitation: G_{sin}/G_{sw} (direct system): columns with planar coil; rows with ring coil.

Clearly, the results for sine wave excitation is higher than square wave excitation, due to square wave can be seen as Fourier series, where the response have several low frequencies components eliminated. Consequently, the result of the sum is lower than the resonance response of the system, when excited by a sine wave.

Fig. 14 shows the curve of the gain ratio for direct system, which they shows that the variation is almost constant in the most cases.

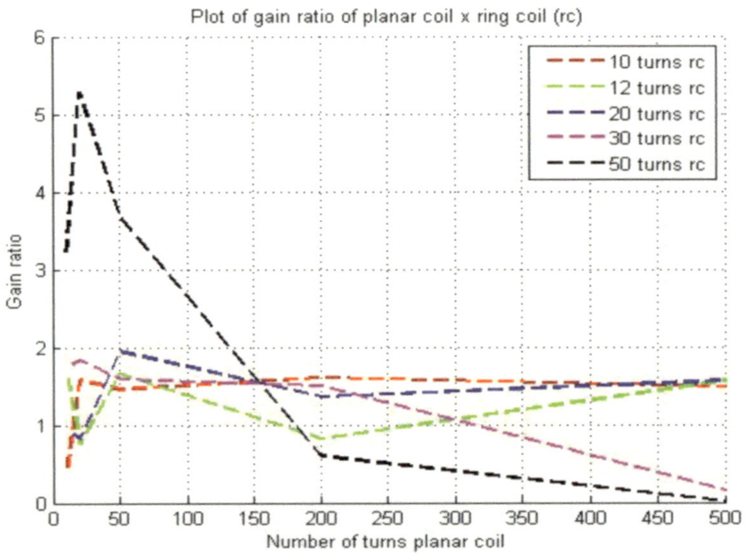

Fig. 14. Graph showing gain ratio for direct system with sine wave excitation (Gsin) and square wave excitation (Gsq): Gsin/Gsq.

2.1.4 Data analysis for excitation with sine wave with inverted system and comparison with square wave excitation

Considering inverted system excited by a sine wave, similar problem is found as inverted system excited by square wave, as well the higher gain of the sine wave excitation.

Considering low frequencies, the response of the system has the behaviour similar as low frequencies in direct system, what we can see in Fig. 15. When frequency increases, behaviour appears in similar way as previously cases, as shown in Fig. 16.

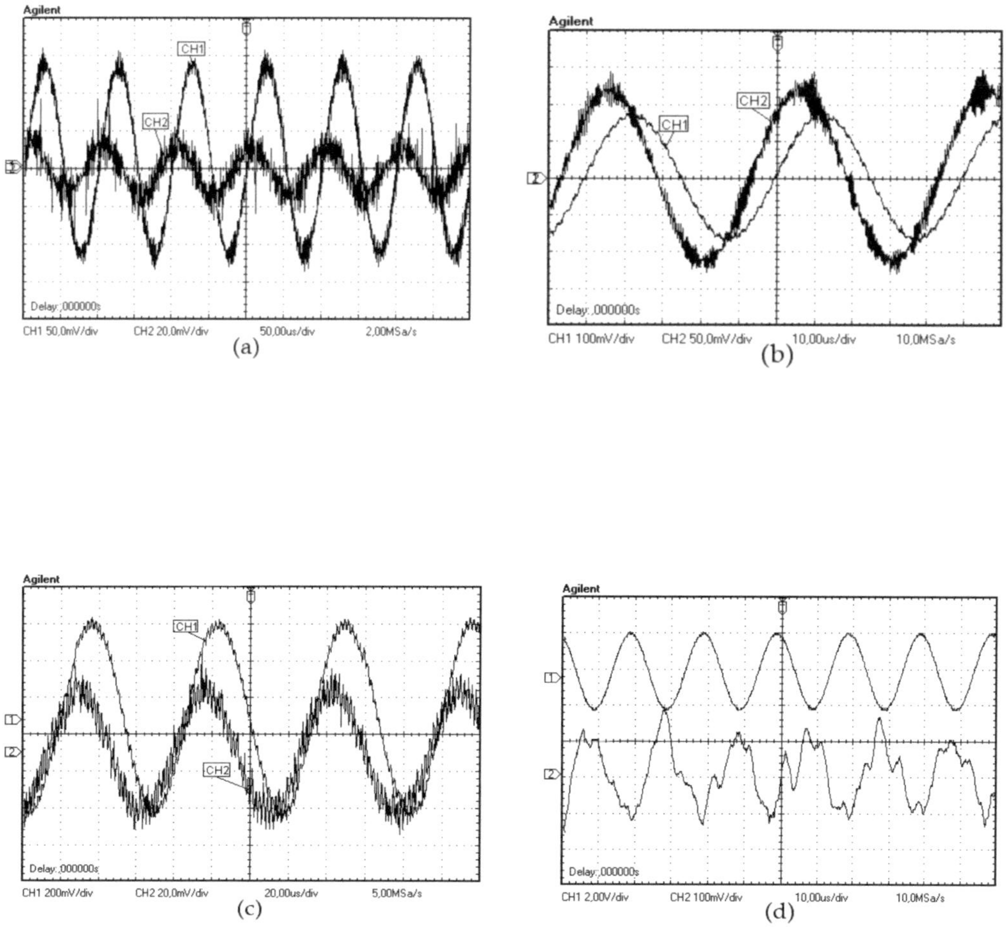

Fig. 15. (a) 10 turns ring coil versus 30 turns planar coil at 5 kHz; (b) 12 turns ring coil versus 200 turns planar foil at 20 kHz; (c) 30 turns ring coil versus 20 turns planar coil at 15 kHz; (d) 50 turns ring coil versus 500 turns planar coil at 50 kHz.

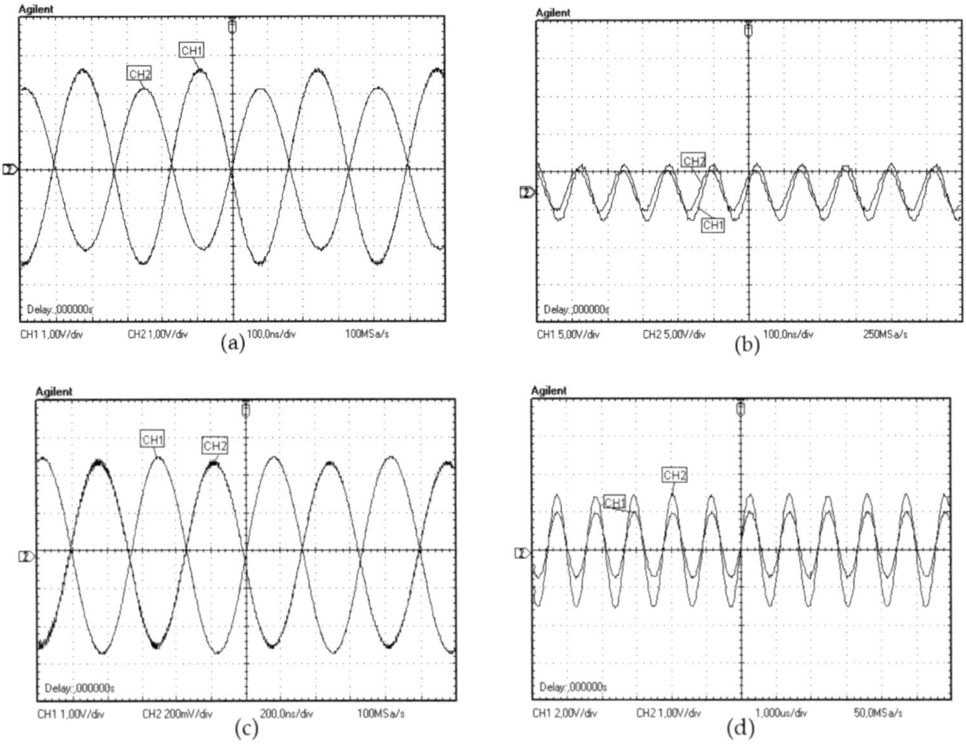

Fig. 16. (a) 10 turns ring coil versus 50 turns planar coil at 3000 kHz; (b) 15 turns ring coil versus 30 turns planar foil at 8000 kHz; (c) 30 turns ring coil versus 500 turns planar coil at 1500 kHz; (d) 50 turns ring coil versus 200 turns planar coil at 900 kHz.

In this case, the problem also presents higher gain, when comparing with square wave excitation. Table 7 shows the gains of some inverted systems, and Table 8 shows the gain ratio of these systems.

Coil	10	12	20	30	50
10	9.68	11.20	14.40	16.96	24.60
20	8.48	8.72	12.00	16.48	19.84
50	4.16	6.08	8.96	8.88	12.48
200	1.22	1.16	2.44	3.84	7.76
500	1.37	0.46	0.71	0.53	0.99

Table 7. Gains of transformers with sinusoidal excitation (inverted system): columns with ring coil; rows with planar coil.

In Fig. 17 are shown the gain ratio of the inverted systems of Table 8. Comparing these curves with direct system, presented in Fig. 14, we can see the similarity with the average gain ratio, where only one system (10 turns ring coil versus 500 turns planar coil) appears as a point out of what is expected.

In this way, both direct system and inverted system, the gain appears almost higher when excited by a sine wave than excited by a square wave. However, in both cases, we can see that considering turn ratio (that is, ideal gain transformer) at resonance, the gain does not satisfy coupled circuit theory. This effect of EMF in secondary transformers when resonance is reached is interesting phenomenon that shows new perspectives for this area. Due to these analyses, several researches can be realized from resonance, applying this special transformer and obtained data and results to new technologies. In this way, researches about induced EMF in cascaded direct systems and inverted systems to reach high output voltages and others are optimal perspectives, as well their applications.

Coil	10	12	20	30	50
10	1.68	1.75	1.76	1.66	1.81
20	1.93	1.73	1.67	1.87	1.94
50	2.42	2.30	1.60	1.21	1.04
200	1.60	1.26	3.18	1.60	1.67
500	43.85	3.87	1.68	0.75	0.58

Table 8. Ratio of the gain transformers with sinusoidal (sin) and square wave (sw) excitation: G_{sin}/G_{sw} (inverted system): columns with ring coil; rows with planar coil.

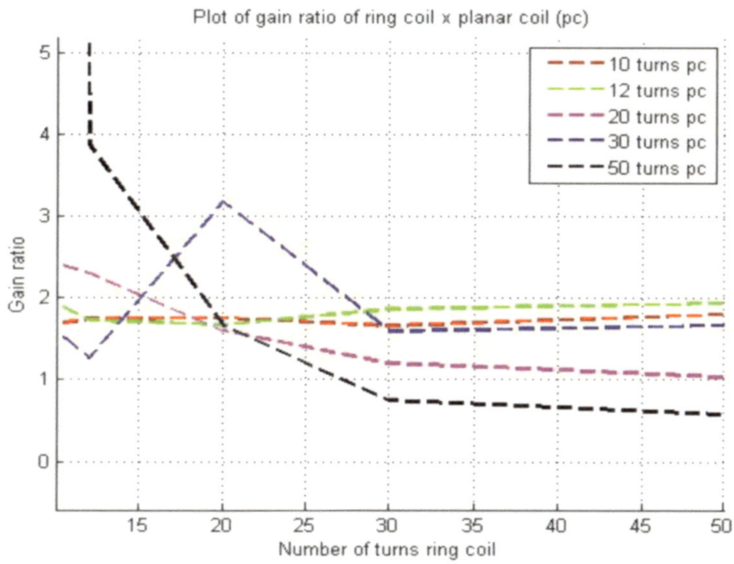

Fig. 17. Graph showing gain ratio for inverted system with sine wave excitation (Gsin) and square wave excitation (Gsq): Gsin/Gsq.

3. Conclusion

This work shows very important results about the induced EMF in coupled circuits (transformers), that not only explains phenomena as high voltage of Tesla transformer, as the found problem of not satisfaction of resonance in circuit theory due to high gain found in output of the special transformers analysed. Some analysis generate simple solutions to this problem, but this work open a new investigative problem in this area, that here is proposed. This work was based on experimental results about air core special transformer, excited by square waves and sine waves in frequencies ranging from 1 kHz to 25 MHz. These transformers were built with planar coils inner ring coils, where initially planar coil was used as primary to verify the induced *emf* response in ring coil and, a posteriori, we invert primary and secondary, exciting ring coil with the square wave, to verify output on planar coil.

In the analysis of the results of the system when excited by a square wave, were observed that the response of the system shows existence of parasitic capacitances, and the response to low frequencies are similar to response of step voltage excitation. But, with the increasing frequency, the responses in each rise and fall of the square wave are added, generating low voltages when this sum of responses are not in phase, and high voltage when the responses are in phase with the square wave, i.e., when is satisfied the relationship $f_r = f_s/n$ (f_r sinusoidal frequency of the response, f_s square wave frequency and n number of cycles of the sinusoidal frequency of the response on semi cycle of the square wave), where this is because energy accumulation in each cycle by the coils in transformer. The higher voltage on output is obtained when the relation $f_r = f_s$ is verified (or $n = 1$). In this case, the maximum values of voltages on output are sinusoidal, showing a resonant response of the system. In both cases (diretc system and inverted system), the response reaches values greater than input, although the turn ratio between coils does not meet the requirements of the circuit theory. So, we observe in results of the inverted system that, when the turn number of planar coil increases too, effects of inductances, parasitic capacitances and resistances generates an active filter on input, which reduces the output voltage. Finally, we see that the better transfer energy observed is obtained to inverted system when turn number ring coil is about 5, and turn number planar coil is great, shown as peak voltage in Fig. 10(b).

When considering sine wave excitation, we note that the system, both direct and inverted sistems, presents higher gain than square wave excitation, that is with average 1.5 times. It is due to amplitude of the sine wave components of the square wave (considering Fourier series), that are lower than peak of the sine wave excitation. The system acts as an filter that eliminates some sine wave components of the square wave, and the response is almost always lower than effect of direct sine wave excitation. Due to results, possibilities of cascaded systems excited by sine wave can generate high resonance voltages, which is shown as new perspectives of application of the high alternating voltages, and others researches with these special transformers, as well induced EMF.

Thus, in both cases, important results are shown, that may be used in researches of electromagnetic interference, computational systems, power electronics, pulse transformers and others excited by square waves and sine waves.

4. References

Anioin, B. A. et al., "Circuit Properties of Coils". IEE Proc.-Sci. Mes. Technol., Vol 144, No. 5, pp. 234-239, September 1997.

Babic, S. I., and Akyel, C.,"Improvement in Calculation of the Self- and Mutual Inductance of Thin-Wall Solenoids and Disk Coils", IEEE Transactions on Magnetics, Vol. 36, No. 4, pp. 1970-1975, July 2000.

Babic, S. I., and Akyel, C., and Kincic, S.,"New and Fast Procedures for Calculating the Mutual Inductance of Coaxial Circular Coils (Circular CoilDisk Coil)", IEEE Transactions on Magnetics, Vol. 38, No. 5, pp. 2367-2369, September 2002.

Babic, S. I., and Akyel, C., "New Analytic-Numerical Solutions for the Mutual Inductance of Two Coaxial Circular Coils With Rectangular Cross Section in Air", IEEE Transactions on Magnetics, Vol. 42, No. 6, pp. 1661-1669, June 2006.

Babic, S. I., and Akyel, C.,"Calculating Mutual Inductance Between Circular Coils With Inclined Axes in Air", IEEE Transactions on Magnetics, Vol. 44, No. 7, pp. 1743-1750, July 2008.

Cheng, K. W. E. et al., "Examination of Square-Wave Modulated Voltage Dip Restorer and Its Harmonics Analysis", IEEE Transactions on Energy Conversion, Vol. 21, No. 3, pp. 759-766, September 2006.

Conway, J. T., "Noncoaxial Inductance Calculations without the Vector Potential for Axissymmetric Coils and Planar Coils", IEEE Transactions on Magnetics, Vol. 44, No. 4, pp. 453-462, April 2008.

Costa, E.M.M. (2009). *Parasitic Capacitances on Planar Coil.* Journal of Electromagnetic Waves and Applications. Vol. 23, pp. 2339-2350, 2009.

Costa, E.M.M. (2009a). *A Basic Analysis About Induced EMF of Planar Coils to Ring Coils.* Progress in Electromagnetics Research B, Vol. 17, pp. 85-100, 2009.

Costa, E.M.M. (2009b). *Resonance on Transformers Excited by Square Waves and Explanation of the High Voltage on Tesla Transformers.* Progress in Electromagnetics Research B, Vol. 18, pp. 205-224, 2009.

Costa, E.M.M. (2009c). *Resonance Between Planar Coils Vs Ring Coils Excited by Sqare Waves.* Progress in Electromagnetics Research B, Vol. 18, pp. 59-81, 2009.

Costa, E.M.M. (2009d). *Responses in Transformers Built Planar Coils Inner Ring Coils Excited by Square Waves.* Progress in Electromagnetics Research B, Vol. 18, pp. 43-58, 2009.

Costa, E.M.M. (2010). *Resonance on Coils Excited by Square Waves: Explaining Tesla Transformer.* IEEE Transactions on Magnetics, Vol. 46, pp. 1186-1192, 2010.

Costa, E.M.M. (2010a). *Planar Transformers Excited by Square Waves* Progress in Electromagnetics Research, Vol. 100, pp. 55-68, 2010.

Huang, Z., Cui, Y., and Xu, W., "Application of Modal Sensitivity for Power System Harmonic Resonance Analysis", IEEE Transactions on Power Systems, Vol. 22, No. 1, pp. 222-231, February 2007.

Hurley, W. G. and Duffy, M. C., "Calculation of Self- and Mutual Impedances in Planar Sandwich Inductors", IEEE Transactions on Magnetics, Vol. 33, No. 3, pp. 2282-2290, May 1997.

Kaware, K., H. Kotama, and K. Shirae, "Planar Inductor", IEEE Transactions on Magnetics, Volume MAG-20, No.5, pp. 1984-1806, September 1984.

Lord, H.W. (1971). *Pulse Transformers.* IEEE Transactions on Magnetics, Vol. 7, pp. 17-28, 1971.

Oshiro, O., Tsujimoto, H., and Shirae, K., "A Novel Miniature Planar Inductor", IEEE Transactions on Magnetics, Volume MAG-23, No.5, pp. 3759-3761, September 1987.

Oshiro O., Tsujimoto, H., and Shirae, K., "Structures and Characteristics of Planar Transformers", IEEE Translation Journal on Magnetics in Japan, Vol. 4, No. 5, pp. 332-338, May, 1989.

Su, Y. P., Liu, X., and Hui, S. Y. R., "Mutual Inductance Calculation of Movable Planar Coils on Parallel Surfaces", IEEE Transactions on Power Electronics, Vol. 24, No. 4, pp. 1115-1124, April 2009.